W9-BIV-407

Shrimps, Lobsters and Crabs

Shrimps, Lobsters and Crabs

Dorothy E. Bliss

COLUMBIA UNIVERSITY PRESS

New York

Columbia University Press Morningside Edition 1990

Columbia University Press
New York   Oxford
Copyright © 1982 Dorothy E. Bliss

Introduction to the Morningside Edition copyright © 1990 Columbia University Press
Reprinted by arrangement with The American Museum of Natural History

Library of Congress Cataloging-in-Publication Data

Bliss, Dorothy E.
  Shrimps, lobsters and crabs / Dorothy E. Bliss.—Morningside
ed.
    p. cm.
    Reprint, with new introd. Originally published: Piscataway, N.J. :
  New Century Publishers, 1982.
    Includes bibliographical references.
    ISBN 0-231-07202-3 (alk. paper).—ISBN 0-231-07203-1 (pbk.)
    1. Decapoda (Crustacea) 1. Title.
  QL444.M33B6   1989
  595.3'84—dc20   89-24031
                                                            CIP

Casebound editions of Columbia University Press books are Smyth-sewn and printed on permanent and durable acid-free paper

Printed in the United States of America

c 10 9 8 7 6 5 4 3 2 1
p 10 9 8 7 6 5 4 3 2 1

Contents

Introduction to the Morningside Edition

During her many years as Curator of Invertebrates at The American Museum of Natural History, Dr. Dorothy E. Bliss included among her duties both preparation of scientific exhibits and responses to queries from the public about identification and habits of crustaceans. Impetus for this book, *Shrimps, Lobsters and Crabs*, was derived initially from a "temporary" exhibit mounted in 1957 (it is still standing), and the idea of collecting and codifying some of the facts most often sought by the public. The book was originally published by New Century Publishers in 1982 and went out of print after about a year. Before her death in December 1987, Dr. Bliss told me that she hoped the book could be reissued in paperback to reach a wider audience.

Although *Shrimps, Lobsters and Crabs* is written in a style accessible to the public, it contains a wealth of technical information about the animals. It is concerned with the crustaceans that most laymen know best, both from the beach and the fish market. Throughout the book, Dr. Bliss has drawn on her own areas of expertise (regeneration, molting, and physiological ecology of land crabs) and has also made use of other experts, ranging from scientific colleagues to commercial fishermen. The book presents a unique overview of these commercially valuable and scientifically important animals that should be of interest to students, beachcombers, and anyone with a scientific curiosity about animals other than themselves. It is a fitting tribute to the memory of Dr. Dorothy E. Bliss and to her concern with crustaceans and their interactions with people.

Linda H. Mantel
City College of New York and
The American Museum of Natural History
August 1989

Preface

Have you ever eaten tomally—or coral—and wondered what part of the animal you were eating? Have you ever wondered why a lobster that is green going into the pot is red when it comes out? Or what the rest of the animal looks like when you buy a frozen rock lobster tail or an iced shrimp tail? Or where these creatures come from? And why?

Do you call a shrimp a prawn and a prawn a shrimp, because shrimp and prawn are the same—or are they?

Have you ever met a crab walking along a path and wondered how the crab has managed to survive so far from the sea, since crabs always live in sea water—or do they?

For all that shrimps, lobsters and crabs are familiar animals, at least on the dinner table, how much do you know about them as moving, feeding, breathing, reproducing organisms—*living* organisms?

This book will provide you with many facts and answer many questions about shrimps, lobsters and crabs. I hope that it will also open new avenues of thought regarding them. This book will introduce you to the ecology of these animals and suggest areas in which their conservation is crucial.

Much of this book concerns the biology of commercially important shrimps, lobsters and crabs. The meat from these animals is high in protein, low in fat, rich in iron and other minerals, and delicate of flavor. It is eaten by many peoples around the world. There is a danger that some marine animals, particularly the larger ones, may be overfished and their numbers may drastically decline. Before restrictions on the fishery existed, this was happening to the Alaskan king crab in the eastern Bering Sea. It may now be happening to large American lobsters on the southeastern part of George's Bank and in areas to the south. To produce a 15-pound king crab or a 30-pound American lobster requires many years. When stocks of such animals are depleted, they can be replenished only with time and effort—if at all.

Most of us realize the need for wise exploitation of natural resources. But to exploit wisely, we should know how animals live, what causes them to thrive and their numbers to multiply, what may be their potential for growth. We should recognize the importance of basic research on these animals, for on such research is advancing knowledge founded.

In short, we should learn all that we can about living forms, and this knowledge should be disseminated as widely as possible. To this end this book is dedicated. But it is also dedicated to the enjoyment of you, the reader. For there is a fascination about shrimps, lobsters, and crabs, their structure, their behavior, the lives that they live. The facts and observations assembled here may help to arouse a continuing interest in these animals and a concern for their future. This is my hope.

                                                              Dorothy E. Bliss

Foreword

Scientific and Popular Names

Every species of animal is given a scientific name, which is formed either from Latin words or from words or combinations of letters that have been Latinized. The scientific name usually consists of two parts, a capitalized italicized generic name and an uncapitalized italicized specific name. The scientific name may be followed by an initial or a proper name that indicates what person described the animal—and sometimes by a date to indicate when the description was published.

For example, take two species of southern marine shrimps belonging to the genus *Penaeus*. The specific name of one shrimp is *aztecus* and of the other *setiferus*. Because in the year 1891 the first of these two species was described and named by the scientist J. E. Ives, its designation in full is *Penaeus aztecus* Ives, 1891 and, in somewhat abbreviated form, *P. aztecus* Ives.

The great Swedish naturalist Linnaeus originally named and described the second of the two species. Therefore, this species is known as *Penaeus setiferus* (Linn., 1767) or *P. setiferus* (L.), either Linn. or the initial L. being generally accepted abbreviations for the name Linnaeus. Note the parentheses around the abbreviated name or initial and the date. These parentheses show that the generic name given to the animal by Linnaeus was something other than the present-day generic name *Penaeus* and that a later worker placed this species in the genus *Penaeus*. Nevertheless, Linnaeus received credit as namer because it was he who first described the animal and gave it a specific name, *setiferus.*

A taxonomist may decide to divide a genus into two or more subgenera. If so, he or she assigns appropriate subgeneric names, which are capitalized, italicized and placed in parentheses between the generic name and the specific name. In 1969, Dr. Isabel Pérez-Farfante, working at the United States National Museum of Natural History, concluded on the basis of certain anatomical features that the genus *Penaeus* should be subdivided into several subgenera and that, accordingly, *Penaeus setiferus* should be known as *Penaeus (Litopenaeus) setiferus* and *Penaeus aztecus* as *Penaeus (Melicertus) aztecus.*

If within a given species a taxonomist decides that there are two or more subspecies, he so indicates by placing after the name of the species a different subspecific name, uncapitalized but italicized, for each subspecies. In her revision of the genus *Penaeus,* Dr. Pérez-Farfante decided that *Penaeus (Melicertus) aztecus,* which occurs along the American Atlantic and Gulf coasts from Martha's Vineyard to Rio de Janeiro, comprises two subspecies, a northern one, *aztecus,* and a southern one, *subtilis.* Thus, the complete designations became *Penaeus (Melicertus) aztecus aztecus* and *Penaeus (Melicertus) aztecus subtilis.*

Taxonomists give a precise scientific label to each animal that they describe so that any scientist using an animal in his studies can specify accurately to his colleagues the identity of the animal. The scientific label also indicates certain relationships regarding the animal, as the examples given above illustrate.

Popular names do not suffice for these purposes. They tell nothing of the relationships between animals, and many popular names may be used for a single animal, the names varying from one locality to another.

The southern marine shrimps of the western Atlantic Ocean provide an excellent example of this. The most frequently used popular name for *Penaeus (Litopenaeus) setiferus* is white shrimp. Yet this animal is also called the common shrimp, grey shrimp, rainbow shrimp, southern shrimp, in Louisiana the lake shrimp, and in North Carolina the green shrimp, green-tailed shrimp, or blue-tailed shrimp. *Penaeus*

(Melicertus) aztecus aztecus is known as the brown shrimp, brownie, green lake shrimp, red-tail shrimp, and in Texas the golden shrimp or red shrimp. *Penaeus (Melicertus) aztecus subtilis* may be referred to as the brown shrimp or dark shrimp, and in South America by a variety of local names. *Penaeus (Melicertus) duorarum duorarum* is called in Florida the pink shrimp, in North Carolina the blue-tailed shrimp or the channel shrimp, and elsewhere the grooved shrimp, brown-spotted shrimp, pink-spotted shrimp, green shrimp, red shrimp, and other names.

Along the west coast of Mexico, Central America, and north–central South America the name white shrimp is applied to *Penaeus occidentalis, Penaeus stylirostris,* and *Penaeus vannamei,* while along the east coast it is applied with equal abandon to *Penaeus (Litopenaeus) setiferus* and *Penaeus (Litopenaeus) schmitti.*

In many other groups of animals the situation regarding popular names is no better. With the spiny lobsters, which are familiar to many readers in the form of frozen lobster tails, the problem has become so acute that Dr. Harold W. Sims, Jr., of the Florida State Board of Conservation has published a short paper entitled: "Let's call the spiny lobster 'spiny lobster.' " For in addition to being known as a rock lobster, langouste, thorny lobster, and long oyster, the spiny lobster is also frequently called a sea crayfish or a marine crawfish. But a close freshwater relative of the American lobster is also called a crayfish or crawfish—or crawdad. And in many parts of the world a freshwater shrimp of the genus *Macrobrachium* is known as a crayfish or crawfish.

We could continue citing examples of this sort almost indefinitely, but these few should suggest the confusion that can arise from the use solely of popular names. When used in conjuction with scientific names, as they are in this book, popular names are a convenience and have a certain value.

Acknowledgments

Many people have helped to make this book a reality, and to all of them I express my appreciation. In the years during which I gathered source material and wrote the book, I incurred heavy indebtedness to colleagues, co-workers, and friends. Probably my greatest debt is to scientists and naturalists from whose published work I have extracted so much information. Some of their publications are cited later in this book in the Select Reading List and the list of Illustration Acknowledgments.

Also great is my indebtedness to members of the staff of the Library at the American Museum of Natural History. Without their patient, cheerful assistance and the Library's extraordinary resources, this book would never have come into being.

For providing me with unpublished information of various sorts, I am grateful to Dr. William Dall, Dr. Horton H. Hobbs, Jr., Dr. Anthony J. Provenzano, Jr., Dr. K. Ranga Rao, and officials at the Virginia Institute of Marine Science, the Massachusetts Division of Marine Fisheries, the United States Fish and Wildlife Service, the National Marine Fisheries Service, and Wakefield/Pacific Pearl Seafoods.

I express my appreciation to Ms. Frederica Leser for making available her color notes on large American lobsters and to Mrs. Francis H. Low for permitting me to cite her observations on penaeid shrimps in Shinnecock Bay.

I am grateful to Mrs. Judith Ann Greenspan for assistance in assembling illustrative material and information on geographical distribution of species, to Mrs. Jane R. Boyer for help in preparing several maps, to Mrs. Rita Due Mogensen and Mrs. Cecilia Ross for typing the manuscript, and to Mrs. Aline Glorieux, Dr. Edwin A. Martinez, Mr. Gerald W. Thurmann, and Mrs. Rose Wadsworth for a variety of helpful services.

I am particularly indebted to Ms. Cady Goldfield for her clear, skillful artwork. My indebtedness extends to members of the Department of Photography of the American Museum of Natural History, who prepared many of the photographic prints used in this book. Credits to persons who took the photographs appear elsewhere, but I wish to note here my appreciation to these persons for permitting use of their material.

For reading the manuscript or portions thereof, I thank Mrs. Jane R. Boyer, Dr. Fenner A. Chace, Jr., Dr. Hélène Charniaux-Cotton, Dr. Penny M. Hopkins, Dr. C. P. Idyll, Dr. Linda H. Mantel, Mr. James K. Page, Jr., Dr. Anthony J. Provenzano, Jr., Ms. Pauline Riordan, Dr. Oswald Roels, and Dr. Harold H. Webber.

I thank Dr. William K. Emerson, former Chairman, and Dr. Ernst Kirsteuer, present Chairman of the Department of Invertebrates at the American Museum of Natural History for their encouragement and support. I also thank Ms. Rosamond Dana and Mr. Douglas Preston for their interest in the book and their concern for its publication. Special thanks go to Mr. Robert W. Hill, Editor-in-Chief, New Century Publishers, Inc., for guiding this book to publication.

It is with gratitude that I acknowledge the help of all of these persons. The content of the book, nonetheless, is my responsibility. If errors exist, they are mine.

Shrimps, Lobsters and Crabs

One

Recognition

Everyone can recognize a shrimp, a lobster, or a crab—or thinks that he or she can. But how many persons are familiar with the structural relationships that exist among shrimps, lobsters and crabs, and their close relatives? Shrimps, lobsters and crabs have been classified with many other animals in the phylum Arthropoda, superclass Crustacea, class Malacostraca, order Decapoda. But on what basis?

Before attempting answers to this question let us make some comparisons between vertebrates and invertebrates. Scientists have described and named over one million species of animals, with only about 6% of them vertebrates. These animals, among which, of course, we number ourselves, have a bony and cartilaginous internal skeleton that includes a backbone. Early in life some vertebrates may have not a backbone, but a long flexible dorsal rod known as a notochord; yet in all except the most primitive vertebrates, the notochord is replaced later in life by a backbone. So significant is the notochord, nevertheless, that the large grouping, or phylum, containing the vertebrates is called Chordata.

Except for several groups of primitive crea-

tures that combine features of both vertebrates and invertebrates, the remaining 94% of living animals constitute the
invertebrates. The prefix "in-" of the word invertebrate
means "without," and all invertebrates are without verte-
brae, that is, without a backbone.

Most present-day biologists classify invertebrates into 27
phyla, of which 10 occur only in the sea and five others mostly
in the sea. Four phyla are without marine forms. Until quite
recently, there appeared to be general agreement that one of
the 27 phyla is Arthropoda and that this phylum includes
insects, spiders, horseshoe crabs, and crustaceans, such as
shrimps, lobsters and crabs. Now, however, there is disa-
greement as to whether a single phylum Arthropoda exists.
Some biologists believe that one evolutionary line has led to
the phylum Uniramia, which includes the myriapods (centipedes, millipedes) and the insects. These biologists would
say that a second evolutionary line has led to a separate
phylum that includes horseshoe crabs, spiders, and
crustaceans—and their close relatives. A few biologists would
separate phyletically the crustaceans from spiders and horseshoe crabs, and possibly even the last two groups from each
other. Thus, according to one's way of thinking, a biologist
can consider the arthropods to constitute one phylum or two,
three, or four phyla. In this book I have preferred to use the
term arthropod as it was originally conceived, to indicate a
member of the phylum Arthropoda.

Recognizing an Arthropod

Taken literally, the term Arthropoda signifies that all members of this group possess jointed appendages. The legs of a
lobster or cockroach, for example, are composed of several
individual parts that meet at movable joints. For contrast,
consider a clam, which belongs to the phylum Mollusca (from
the Latin word *mollis*, meaning "soft," in reference to the soft
body within the shell). Obviously, a clam has no jointed
appendages and can never belong to the phylum Arthropoda.

An advantage possessed by most flexible-legged arthropods over most mollusks—and for that matter over the

members of most other invertebrate phyla—is their maneuverability. In this regard, nevertheless, the arthropods are more than matched by certain highly developed, swimming marine mollusks known as cephalopods. Among existing species of cephalopods, the pearly nautilus alone retains an external shell. Internal vestiges of a shell are present in the squid (as the pen) and the cuttlefish (as the cuttlebone), but not in the octopus, which lacks all traces of such a structure. The fragile "shell" of the paper nautilus, *Argonauta,* is something very different, being a boat-like structure that the tentacles of the female secrete for carrying eggs.

Shrimps, lobsters and crabs share several other characteristics with their fellow members of the phylum Arthropoda. First, with the exception of some crabs known as anomurans, they are bilaterally symmetrical; that is, the two longitudinal halves of the body are essentially alike. A contrasting type is radial symmetry, of which the starfish of the phylum Echinodermata (literally "spiny skinned") is a notable example. In the starfish several "arms" radiate from a common center.

Secondly, all arthropods have a hard outer covering that is technically termed an exoskeleton and is popularly known as a shell. This structure is composed of (1) a protein that has become chemically changed or "hardened," (2) a resistant complex chemical compound named chitin (pronounced KȲ-tǐn), and in some forms (3) compounds of calcium. The exoskeleton covers the entire outer surface of an arthropod's body and even lines the anterior (frontal) and posterior (rear) regions of its digestive tract as well. Yet because of pliable, noncalcified membranes of chitin and unhardened protein between their several sections, the appendages remain flexible at each joint.

Due to the hard exoskeleton in which arthropods are encased, these animals can increase in size only periodically—during the recurrent but brief hours that follow the casting off of an old shell and precede the hardening of a new one. Enlargement of body size that occurs at this time is due to heightened internal pressure. This, in turn, is due to a rapid uptake of water or, as in many terrestrial arthropods, of air.

The actual shedding of the old shell is called ecdysis (ĕk-DȲ-sĭs), after the Greek word *ekdysis,* meaning "a getting out." Ecdysis is preceded and followed by much metabolic activity, during which the new exoskeleton is formed, missing limbs are regenerated, and in shrimps, lobsters, crabs, and other calcified arthropods, the old shell is selectively decalcified and the new one calcified. To this entire complex of activities that constitute growth in an arthropod, the term molt is applied.

In common with other arthropods, all shrimps, lobsters and crabs have bodies that are segmented; that is, they are composed of linearly repeating sections. The total number of segments in the arthropod body varies widely. Certain species of millipedes have more than 100, whereas an adult grasshopper has only 20. In general, primitive arthropods tend to have many segments, more advanced arthropods few.

Some segments bear one or two pairs of appendages. Indeed, in the centipedes almost every segment carries one pair of walking legs. However, among less primitive arthropods usually fewer segments bear appendages—and the appendages that are present tend to be highly modified for specific functions. In shrimps, lobsters and crabs, for instance, appendages of the head and the anterior portion of the thorax, the middle of the three chief divisions of the crustacean body, are modified for detecting, getting, holding, and chewing food, those of more posterior regions of the thorax for walking and in certain crabs for swimming, and those of the tail for mating and egg-holding and in shrimps—but not in lobsters and crabs—for swimming.

The basic segmentation of the arthropod body extends to the nervous system, which is built on the same plan as is the nervous system of the segmented worms (phylum Annelida). In annelids such as, for example, the earthworm, there is one pair of fused ganglia in virtually every segment. Since a ganglion is made up of the massed bodies of nerve cells, this means that there is one local center of nervous integration in almost every section of the annelid body.

Essentially the same can be said of the arthropod nervous

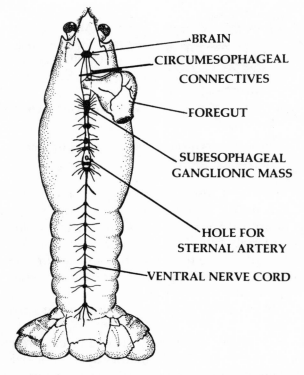

BRAIN

CIRCUMESOPHAGEAL
CONNECTIVES

FOREGUT

SUBESOPHAGEAL
GANGLIONIC MASS

HOLE FOR
STERNAL ARTERY

VENTRAL NERVE CORD

Fig. 1. The central nervous system of a crayfish.

system. However, in most species of arthropods fusion of
ganglia has led to fewer such centers in correspondingly
fewer body segments. Thus, an arthropod such as a freshwater crayfish or its close relative, the American lobster, has
(1) a dorsal brain formed from several pairs of fused ganglia in
the head and connected by means of two nerves (circumesophageal connectives) with (2) a ventral mass of fused
ganglia known as the subesophageal ganglionic mass, and (3)
a paired, but fused, ventral nerve cord that extends posteriorly the length of the body and is enlarged at intervals
where paired fused ganglia occur. Even more fusion of ganglia has occurred in a crab.

Shrimps, lobsters and crabs, and all other arthropods, have
a dorsally situated heart. In its most primitive form, the ar-

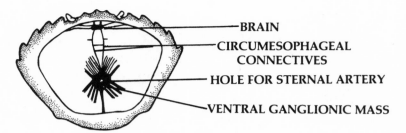

BRAIN
CIRCUMESOPHAGEAL
CONNECTIVES
HOLE FOR STERNAL ARTERY
VENTRAL GANGLIONIC MASS

Fig. 2. The central nervous system of a crab.

thropod heart is a muscular tube lying above the digestive tract and running almost the entire length of the body. Among more advanced arthropods (for example, shrimps, lobsters and crabs) the heart is a compact single-chambered sac that lies within the thorax. Whether tube or sac, the heart has in its walls several openings known as ostia. Through these ostia blood enters the heart from the pericardium, which is a blood-filled chamber, or sinus, that surrounds the heart.

In contrast with its "closed" form in vertebrates (and in cephalopod mollusks) the circulatory system of arthropods is "open." This means that although elastic arteries and thin-walled elastic capillaries occur in many species of arthropods, there are no veins. Instead, the blood returns to the heart by way of interconnecting spaces known as venous sinuses, which communicate with the percardium.

In order to keep blood flowing steadily through the body of an arthropod, an open circulatory system depends upon a head of hydrostatic pressure that is established principally by contractions of the beating heart. The arteries contain almost no muscle. Yet in shrimps, lobsters and crabs "accessory hearts" may help to circulate the blood. These are blood vessels that can decrease in volume by the contraction or increase in volume by the relaxation of muscles running either through or closely beside the blood vessels.

Vertebrates such as ourselves have two circulating substances, one known as blood and the other as lymph. Blood consists of cells plus a fluid medium called plasma and is

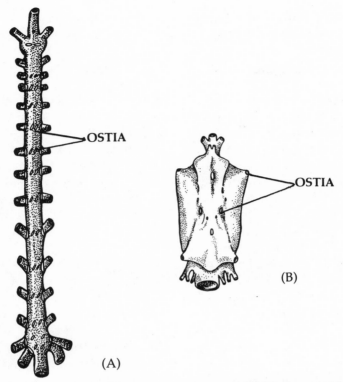

Fig. 3. (A) Long, tubular heart of a stomatopod. (B) Compact heart of a crayfish.

confined to the blood vessels. Lymph (literally "clear water") is a colorless substance derived from blood. It consists in part of plasma that has passed through the walls of capillaries into the tissue spaces and bathes the tissues.

As a consequence of its open circulatory system, an arthropod does not have two separate substances known as blood and lymph. What we have been calling the blood of arthropods is not only a circulating material composed of cells and plasma and therefore equivalent to the blood of vertebrates, but also a tissue-bathing substance analogous to lymph. Hence this substance is more accurately termed hemolymph, the first part of this word derived from a Greek word meaning "blood." The extensive spaces of the ar-

thropod body through which hemolymph circulates are called collectively the hemocoel, or literally "blood hollow."

Recognizing a Crustacean

Crustaceans, including shrimps, lobsters and crabs, as well
as millipedes, centipedes and insects, possess a pair of cutting
or crushing jaws known as mandibles. In structure, the mandibles of crustaceans differ from those of millipedes, centipedes, and insects. All of these groups, nonetheless, have
frequently been included in the term mandibulate arthropods. Because pincer-like first appendages known as
chelicerae occur in spiders, sea spiders, scorpions, mites,
ticks and horseshoe crabs, these animals have usually been
called chelicerate arthropods.

Of the mandibulate arthropods, the presence of one pair of
feelers, or antennae, distinguishes the millipedes, centipedes
and insects. Shrimps, lobsters and crabs have *two* pairs of
antennae, and these are a key diagnostic feature of the superclass Crustacea (the term Crustacea is derived from a Latin
word meaning "crust," with reference to the hard shell).

Another noteworthy feature of *some* members of the Crustacea is that they have a covering known as a carapace, meaning shell. A carapace is lacking in other crustaceans, such as
fairy shrimps, copepods and certain primitive freshwater
crustaceans like *Anaspides*. It is also lacking in isopods such as
the pill bugs and sow bugs (wood lice) and in amphipods,
which include the swiftly swimming *Gammarus* that occurs in
tide pools and in freshwater ponds.

When present, a crustacean carapace may assume any one
of many forms. In water fleas, of which *Daphnia* is an example
familiar to students of elementary biology, the carapace is a
bivalve shell that does not cover the head. In clam shrimps
and ostracods the carapace is likewise a bivalve shell, but in
these forms it covers almost the entire body, including the
head.

In barnacles, which, despite their aberrant appearance, are
also crustaceans, the carapace is a fleshy mantle. Except for
stalked forms like the goose barnacle *Lepas*, in which an ex-

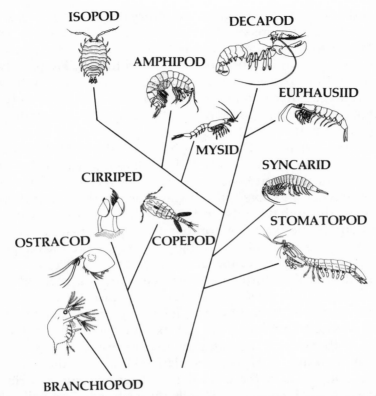

ISOPOD

DECAPOD

AMPHIPOD

EUPHAUSIID

MYSID

SYNCARID

CIRRIPED

STOMATOPOD

OSTRACOD

COPEPOD

BRANCHIOPOD

Fig. 4. Eleven major types of crustaceans, with possible interrelationships indicated by branching lines.

tension of the head protrudes in the form of a stalk, the entire body of a barnacle is covered by this fleshy mantle. Embedded within the mantle and strengthening it are calcified plates, which constitute the shell and are responsible for the extreme hardness and roughness that we associate with barnacles.

The carapace in shrimps, lobsters and crabs and in the small pelagic, shrimp-like crustaceans known as mysids and euphausiids is shield-like and often fused with some or all segments of the thorax. The carapace of these forms provides protection for the important anterior region of the body, with its numerous vital organs that include the heart, digestive glands, reproductive organs, excretory organs and others. In some cases, for instance, in shrimps, lobsters, crabs and euphausiids, but not in mysids, the shield-like carapace obscures the boundary between head and thorax. Frequently a groove, the so-called cervical groove, is the only clue to the existence of this boundary. In such instances, the body is often said to be composed of only two regions, cephalothorax and abdomen.

Respiration in the Crustacea generally takes place by means of gills, although in very small crustaceans respiration can occur over the entire body surface. As water flows over the gills, oxygen dissolved in the water enters the blood that is passing through the gills, and carbon dioxide leaves the blood and enters the water. A comparable type of respiratory exchange goes on within our lungs, although here the oxygen-bearing medium is air rather than water.

In its simplest form the crustacean gill is a plate-like or bladder-like sac containing many capillaries. Amphipods have this type of gill. Shrimps, lobsters and crabs have more complex gills: within a central axis of each gill are blood vessels that supply blood for the respiratory exchange taking place in numerous highly vascularized branches that emerge from both sides of the central axis. Theoretically, a shrimp, lobster or crab can have 32 gills on each side. Actually the largest number (in a species of shrimp) is 24 on a side. At the other extreme, the crab *Pinnotheres,* which lives commensally within the shell of mussels and oysters, has only three gills on each side.

Strictly speaking, the gills of a shrimp, lobster or crab lie outside of the body of the animal. Yet they are situated entirely within a chamber, the branchial chamber, of which there is one on each side and which is formed from a deep lateral fold of the carapace. Access to each branchial chamber exists through anterior or posterior openings or through openings just above the legs.

In most crustaceans gills occur on or close to the base of the thoracic appendages. But in stomatopods branching tufts of respiratory filaments arise from the base of the abdominal appendages. These abdominal gills are believed to supplement several pairs of gills on the anterior thoracic appendages. In isopods, the abdominal appendages (one pair or two pairs or all of them) constitute the sole respiratory organs. In both stomatopods and isopods, the existence of abdominal gills is associated with a long tubular heart, which extends well into the abdomen.

A feature of most (but not all) crustaceans is their compound eyes, so-called because they are made of many parts, or ommatidia. Each ommatidium has a transparent covering known as a facet and, beneath the facet, a corneal lens, through which light enters. Underneath the corneal lens is a prism termed the crystalline cone, which funnels light proximally, and, except in crabs, an extension of the cone called the crystalline tract.

At the proximal end of the crystalline cone (or crystalline tract) is the retinula, a group of seven sensory nerve cells arranged radially. These cells surround a rod-like refractile body, the rhabdom, which contains visual pigments and is sensitive to light. The rhabdom consists of many layers of minute, inwardly projecting tubular extensions of the retinula cells, with tubules of alternate layers lying at right angles to one another. Fibers from the retinula cells pass through a

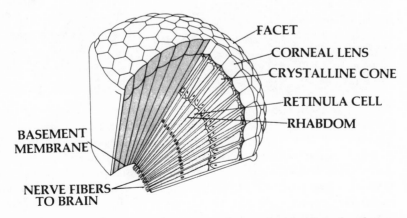

Fig. 5. Sectional diagram of compound eye of crab, showing internal structure of eight of numerous ommatidia that make up this type of eye. (After Sherman and Sherman)

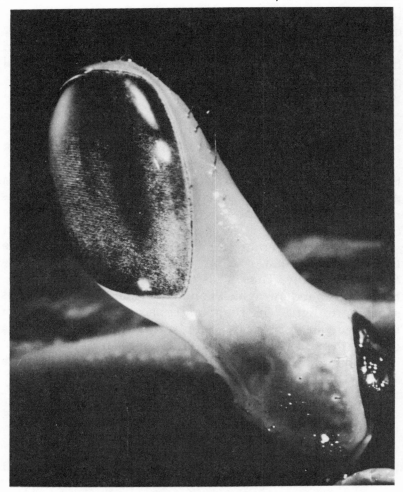

Fig. 6. The right compound eye of the large land crab *Cardisoma guanhumi*. The eyestalk is in upright position and is being observed from the front. Thousands of tiny hexagonal facets are visible, regularly arranged in horizontal rows. Beneath each facet is a minute two-channel analyser for the perception of polarized light. (Courtesy of Talbot H. Waterman and Kenneth W. Horch. From: Mechanisms of Polarized Light Perception, *Science,* vol. 154, pp. 467–475, 1966)

basement membrane, which holds the fibers in a fixed position relative to the surface of the lens. The fibers then join to form the optic nerve, which carries light-generated nerve impulses into the central nervous system.

In bright light, dark pigment granules surround and screen the crystalline cone and rhabdom, thus reducing the amount of light reaching the rhabdom. In dim light the granules concentrate into two small masses, leaving the crystalline cone and rhabdom exposed and thus enhancing the light-gathering power of the ommatidium.

The ommatidia of crustacean eyes vary in number, from a few (20 to 25 in the pill bug *Armadillidium*) to several thousand (3000 in the stomatopod *Squilla mantis*) and to an estimated 14,000 in the American lobster, *Homarus americanus*. Each ommatidium forms an erect image, with an erect mosaic, the total image formed by all ommatidia. The greater the number of ommatidia, the closer the total image agrees with the object viewed.

Because moving objects affect a series of ommatidia in succession, compound eyes can readily perceive movement, and it is believed that they can perceive depth and distinguish between colors as well. Compound eyes can also detect linearly polarized light.

Two Italian scientists, Drs. L. Pardi and F. Papi, discovered that if skies are clear, the amphipod *Talitrus saltator* and the isopod *Tylos latreillei* can orient correctly when in shade, even though these crustaceans ordinarily use the sun or moon to determine compass direction. If clouds cover the sky, however, the animals react in a disoriented manner. The scientists concluded that in the shade the animals orient to the light of an unclouded blue sky; this light is linearly polarized in patterns that depend mainly on the position of the sun.

In more recently published work on the crab *Cardisoma guanhumi*, Dr. Talbot H. Waterman of Yale University showed that the rhabdom of an ommatidium in a compound eye is a two-channel analyzer, in which the two directions of maximum sensitivity to polarized light correspond to the two axes, vertical and horizontal, along which the minute tubules of the rhabdom are arranged. These, in turn, correspond to the vertical and horizontal axes of the eye when the eye is up in a seeing position.

Amphipods, isopods and crabs are not unique in their capacity to orient to polarized light. More than 90 species of animals, largely arthropods and cephalopod mollusks, are able to detect linearly polarized light and to determine its plane of polarization.

Compound eyes of the crustacean type occur in insects and in the horseshoe crab, *Limulus polyphemus*, but not in the remaining chelicerate arthropods or in millipedes and centipedes. The latter three groups possess simple eyes, or ocelli, each of which in its least complex form corresponds to a single ommatidium of a compound eye. More elaborate ocelli occur in many insects.

Among crustaceans with compound eyes, many groups, including shrimps, lobsters, crabs, euphausiids, mysids and stomatopods, bear them on movable stalks. On the other hand, compound eyes are nonstalked, or sessile, in isopods and amphipods. Copepods lack compound eyes, as do barnacles when adult.

Recognizing a Decapod Crustacean: Shrimp, Lobster or Crab

Shrimps, lobsters and crabs belong to the class Malacostraca, in which, typically, the compound eyes are on stalks and there are 19 segments in the body, five in the head, eight in the thorax, and six in the abdomen. In addition, the first pair of antennae, called the antennules, are often biramous; that is, they have two branches. Furthermore, in the Malacostraca, reproductive openings of the female are on the sixth thoracic segment, while those of the male are on the eighth thoracic segment. Within the class Malacostraca, 14 orders are now generally recognized, with shrimps, lobsters and crabs constituting the order Decapoda.

The name Decapoda (from the Greek, meaning "ten feet") suggests the principal diagnostic feature of this order. With relatively few exceptions, all crustaceans in the order have five pairs of legs, that is, 10 legs in all. This is true of no other order of the class Malacostraca. Stomatopods, for example, have three pairs of legs, isopods and amphipods have seven pairs. Not all five pairs of legs of decapod crustaceans are necessarily used for walking. For instance, in the American lobster the first pair is modified as very large claws, used for seizing, cutting and crushing. In the blue crab, *Callinectes sapidus*, the last pair is paddle-shaped and is used for swimming.

At first glance, members of the order Decapoda would appear readily divisible into two groups of animals, those with long tails and those with no tails. But the apparently tailless group does in fact have a short tail, which it carries tucked underneath the body. This short-tailed group is a large one and is known as the infraorder Brachyura, after the Greek words *brachys*, meaning "short" and *oura*, meaning "tail." The Brachyura, which in popular terms are the true crabs, include such well-known forms as the blue crab and the green crab, *Carcinus maenas*. Scientists differentiate true crabs from anomuran crabs, which form another part of the Crustacea Decapoda known as the infraorder Anomura.

More than anything else, the fact that all crabs carry the tail tucked underneath the body is what makes a crab look like a crab—and both true crabs and most anomuran crabs have this in common. Yet in true crabs all five pairs of thoracic legs are well developed, whereas in anomuran crabs the fifth pair of legs is greatly reduced in size. As a result, anomuran crabs often look as if they have only four pairs of legs. In some species of anomuran crabs the very small fifth pair of legs is carried hidden within the branchial chambers and becomes visible only when the exoskeleton covering the gills is dissected away. Furthermore, the sixth abdominal segment of anomuran crabs bears two appendages (uropods). True crabs lack these appendages.

Probably the best known representative of anomuran crabs is the king crab, *Paralithodes camtschatica*, which is a major source of canned and frozen crab meat in the United States. Less well known among anomuran crabs are the porcellanids, or porcellain crabs, which have a light, brittle shell and are able to cast off (autotomize) an injured leg with amazing speed.

A female king crab has a particularly striking feature, the asymmetrical way in which calcified plates are arranged on

Fig. 7. Top: In this large anomuran crab, the king crab, *Paralithodes camtschatica*, the last (fifth) pair of thoracic legs is very small and kept hidden within the branchial chambers. A portion of the shell of this specimen has been cut away, and the right fifth leg and the most posterior gills are now visible. The piece of shell that was removed lies at the right. An enlarged view of the cut-away area appears below. Center: The external (left) and internal (right) surfaces of the abdomen of a female king crab. Note the asymmetrical arrangement of calcified plates on the external surface and of abdominal appendages (pleopods) on the internal surface. This asymmetry suggests the close relationship between king crabs and hermit crabs. (Courtesy of the American Museum of Natural History)

her abdomen. Furthermore, in female king crabs, and in female hermit crabs, abdominal appendages occur only on the left side. This asymmetry suggest a close relationship between king crabs and hermit crabs and is responsible for their name, Anomura, which is based on Greek words that mean "irregular tailed." In the male king crab the abdo-

(A)

Fig. 8. (A) A female hermit crab, showing abdominal appendages, which occur only on one side, and the roughened file-like surfaces (arrows) at the tip of the fourth and fifth thoracic legs and of the terminal appendages. By pressing against the inside of the shell, these roughened areas steady the body of the hermit crab in the shell. (Courtesy of Anthony J. Provenzano, Jr.) (B) Impressionistic view of hermit crab as it would occupy the castoff shell of a univalve mollusk. (Based on original drawing by J. A. Allen)

men lacks appendages and the calcified plates are arranged symmetrically.

Hermit crabs are odd creatures, with a typically hard cephalothorax but with a very soft, twisted abdomen that the

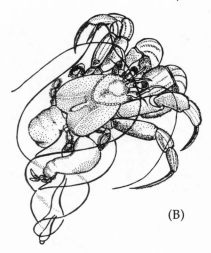

(B)

animal keeps hidden within a discarded snail shell. The terminal appendages, or uropods, of a hermit crab have roughened, file-like surfaces that the animal presses outward against the walls of the shell. Assistance in this holding operation is provided by the fourth and fifth pairs of legs, each of which has a roughened area at its tip that presses against the shell and helps to steady the body of the hermit crab in the shell.

The infraorder Anomura also includes the galatheids, which are a group of long-tailed, lobster-like animals that are often referred to as "squat lobsters." Galatheids can be distinguished from lobsters by the way they hold the abdomen, keeping it partly bent underneath the cephalothorax. The last pair of legs, which are small and slender, they may keep folded within the branchial chambers.

Galatheids comprise some 230 described species, of which two species, *Pleuroncodes planipes* and the subantarctic *Munida gregaria*, can be seen as adults in swarms at the surface of the ocean. *Pleuroncodes planipes* is often called the pelagic red crab, or simply the red crab, even though it is not particularly crab-like in appearance. This brilliantly colored galatheid lives partly on the bottom of the ocean and partly at the surface during its first two years of adult life. After the second

Fig. 9. A pelagic red crab, one of thousands washed ashore at Monterey Bay, California, in March 1973. (Courtesy of Ralph Buchsbaum)

year, pelagic red crabs are strictly benthic; that is, they remain on the bottom of the ocean, separating themselves from younger adults by moving in large masses into deeper water.

While in the surface-swimming phase, pelagic red crabs are found on the continental shelf off the western shore of southern Baja California, mainly south of Punta San Eugenio. But the complex system of oceanic currents that exists in this region (see Chapter 2) causes some red crabs to be swept away from coastal waters and well out into the Equatorial Current. Others, caught up in northward flowing countercurrents, may be carried as far north as Monterey.

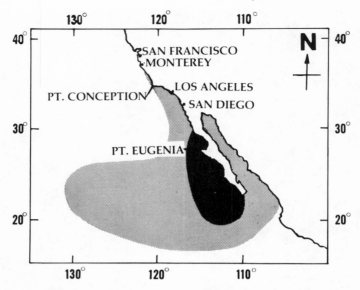

Fig. 10. Distribution of pelagic red crab. Darkly shaded area indicates region of greatest abundance. Westerly bulge reflects south-
westerly sweep of California Current off Baja California. (After
Boyd)

Countless numbers of red crabs can sometimes be seen
swimming at or near the surface of the ocean. Here they feed
on a single-celled green plant known as a diatom and on a
single-celled animal called a dinoflagellate, both of which are
abundant because of upwelling and the nutrients that upwelling brings from the ocean's depths. Pelagic red crabs can be
so numerous that a traveler once reported his ship appeared
to crunch through them for at least 10 miles. Occasionally
swarms of these crabs are washed up on the beach. Such
strandings have occurred at Monterey and San Diego, and
they are common south of Punta San Eugenio.

The pelagic red crab, *Pleuroncodes planipes*, is an important
item of food for large fishes, including the yellow fin tuna,
skipjack tuna, and albacore. Several species of birds and
certain aquatic mammals, such as the grey whale and the sea
lion, also feed on the red crab. A close relative, *Pleuroncodes
monodon*, is a *langostino* that occurs off the coast of Chile; it,

too, is often called the red crab. This galatheid, which as an adult appears to be strictly benthic, with no pelagic phase, occurs in large concentrations on the bottom of the ocean and constitutes a significant part of the diet of fishes, including several large species of commercial importance. In addition, *Pleuroncodes monodon,* along with another strictly benthic galatheid, *Cervimunida johni,* serves as the base of a fishery that is a part of the Chilean shrimp fishery. The product of this fishery for galatheids is largely exported as frozen tails.

Lobsters and their freshwater relatives, the crayfishes, are decapod crustaceans very different from crabs. Because of their long tails, lobsters and crayfishes have often been referred to as macrurans, after the Greek words *makros:* long and *oura:* tail. In general, there are two kinds of lobsters: true lobsters of the infraorder Astacidea; and spiny lobsters, or rock lobsters, of the infraorder Palinura. A true lobster, such as the American lobster, has two large claws and a stiff tail fan. In contrast, a spiny lobster, or rock lobster, lacks large claws and has a flexible, leathery tail fan. Examples of the latter are the Florida, or Caribbean, spiny lobster, *Panulirus argus,* and the South African rock lobster, *Jasus lalandii.*

Among shrimps, the most primitive are the penaeids, which form the backbone of the shrimp fishing industry in North and South America and in many parts of Asia. In a number of ways, the penaeid shrimps illustrate the generalized decapod crustacean condition and thus are extremely valuable in the study of crustacean morphology. The anatomy of the shrimp *Penaeus setiferus* is discussed in Chapter 4.

Structurally, shrimps of the family Sergestidae are more highly specialized, with the last pair of legs reduced in size or lacking, and the gills few or absent. An example of a sergestid shrimp without gills is *Lucifer faxoni,* a small, thin creature with a greatly elongated cephalothorax that places the eyes, antennules, and antennae far in front of the mouth. As with many other sergestids, the primary home of this species is the ocean, from the surface to a depth of 300 feet, but the shrimp also occurs in estuaries and has been observed in swarms outside of the harbor at Beaufort, North Carolina.

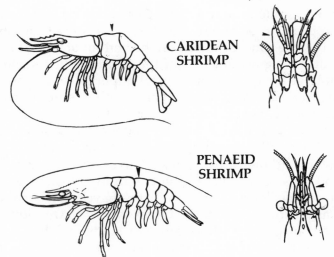

**Fig. 11. How to distinguish caridean shrimp from penaeid shrimp.
In caridean shrimp, back usually forms sharp angle; in penaeid
shrimp, back forms smooth curve. In caridean shrimp, lower part
of second abdominal segment (arrow) overlaps adjacent segments;
in penaeid shrimp, this is not so (arrow). In caridean shrimp,
antennal scale (arrow, upper right) is relatively large; in penaeid
shrimp, antennal scale (arrow, lower right) is relatively small.**

Shrimps of the infraorder Caridea include the freshwater
genus *Macrobrachium,* literally "long-armed," in which the
second pair of thoracic legs is extremely elongated and terminates in a small claw. The infraorder Caridea also includes the
marine and freshwater genus *Palaemonetes* and the marine
genera *Crangon* and *Palaemon.*

Among the most unusual carideans are snapping shrimps,
or pistol shrimps, of the family Alpheidae. These animals
have one claw much enlarged. A prominent tubercle, or
"tooth," on the movable "finger" of the large claw fits snugly
into a depression on the immovable "finger" in much the way
a peg fits into a hole. When the claw closes suddenly, there
results a sharp cracking sound as the two "fingers" come
forcibly together. In addition, the entry of the "tooth" into the
depression results in the emission of a jet of water, which may

help to drive off an attacking enemy. A large mass of snapping shrimps is reported to sound like popcorn popping.

Two families of caridean shrimps, the Palaemonidae and the Hippolytidae, as well as the family Stenopodidae of the infraorder Stenopodidea, include species known popularly as cleaner shrimps. These are small, colorful creatures that establish "cleaning stations" on rocks and coral reefs. Standing at prominent vantage points, the shrimps wave their brightly colored antennae and sway their body back and forth. In response, fishes move up to the shrimps, extending their fins and opening their mouth and gill covers. The shrimps move rapidly over the body of each fish, removing parasites and eating food particles that remain on the teeth and gills. Eventually, their hunger presumably satisfied, the shrimps depart, often leaving many fishes still waiting in line to be cleaned.

You may often have wondered about the technical differences between shrimps and prawns. According to British usage, the term shrimp is restricted to forms having a crangonid-like appearance, while prawn is used for forms

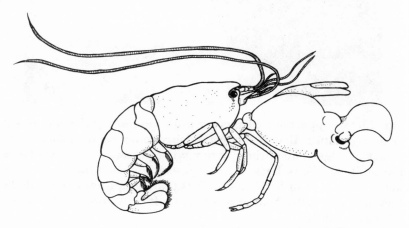

Fig. 12. Snapping shrimp, showing tubercle, or "tooth," on movable part of large claw and depression into which tubercle fits. Rostrum and eye, drawn clearly here, are often indistinct or even invisible when snapping shrimp is viewed from side. (After Schmitt)

Fig. 13. A peppermint cleaner shrimp, *Lysmata grabhami,* at the cleaning station that it has established on a small promontory under a ledge. The shrimp is removing parasitic copepods from the body of a fairy bass, *Anthias squamipinnis.* Other fairy bass await their turn to be cleaned. This photograph was taken in 20 feet of water at the Nature Reserve, Gulf of Aqaba, Eilat, Israel. (Douglas Faulkner)

with a palaemonid-like structure. The American taxonomist Dr. Fenner A. Chace, Jr., of the United States National Museum of Natural History, has pointed out that this distinction works only for species restricted to northern Europe.

When applied to forms found elsewhere, the distinction is artificial. Thus, according to the British definition, penaeids would have to be called prawns, but in the United States at least, they are always referred to as shrimps. According to the British definition, the only true shrimps in North America are the crangonids of the Pacific coast—and indeed, with one exception, these are called shrimps. Yet the exception, the largest species of commercial importance, is often called a prawn on the basis of its relatively large size. Thus, American usage bestows the term prawn upon species of large size.

A solution, as adopted by Dr. Chace, is to use the term prawn only for noncrangonid types. This is, however, a taxonomist's solution. For the layman and for the fisherman, in the United States at least, the term shrimp will no doubt continue to be used for small species and the term prawn—or jumbo shrimp—for large ones.

In this chapter we have seen how shrimps, lobsters and crabs (and other decapod crustaceans) differ from one another, from their fellow crustaceans, and from other arthropods. Before looking more carefully at the structure of shrimps, lobsters and crabs and noting how their structure relates to function, we shall first note the geographical areas in which these animals are found, often in numbers large enough to support a fishery. We shall also seek to determine why these concentrations of animals exist.

Two

Oceanic Distribution
and Fisheries

Living organisms evolved within the ocean, and sea water represents the most benign of any external milieu. Thus, it is fitting that we start our overview by considering the distribution and fisheries of shrimps, lobsters and crabs that live in the ocean. Yet we cannot do this without examining the great oceanic currents. For, to a large extent, these currents, either directly or indirectly, determine where marine animals are to be found.

Oceanic Currents Along the North American West Coast

On the coast of southern California, there is a headland known as Point Conception. A northbound traveler, rounding Point Conception, may experience a drop in the temperature of the air; and, should this person test the surf pounding the shore, he or she would find that the water also has become much cooler.

One late December, making such a trip by train, I watched swimmers and surfboarders frolic south of Point Conception and then, within the hour, saw people clad in heavy clothing and bundled against a cold wind striding along the beaches north of the Point. A change also was evident in the vegetation, with palm trees and cacti giving way to bushes and trees characteristic of more northern regions of California.

The primary cause of this abrupt change of climate is the oceanic currents that run parallel to the shoreline of California. One, carrying cool subarctic water southward, is the California Current. The other, bringing warm tropical water northward, is a deep undercurrent from Baja California. In late fall and early winter, when north winds are weak or absent, the undercurrent surfaces well inshore of the main stream of the California Current. Here it becomes known as the Davidson Countercurrent from Point Conception northward to Oregon and Washington and as the Southern California Countercurrent south of Point Conception. The latter countercurrent expands into a spiral or circle, known as a gyre, inside the islands off southern California and then swings northward around Point Conception, where it joins the Davidson Countercurrent.

Both the Davidson and the Southern California Countercurrents lie very close to shore. But because the Southern California Countercurrent is weak, its waters remain off the coast of southern California for some time, becoming noticeably warmer than offshore waters and those of the Davidson Countercurrent north of Point Conception. In November, for example, the average temperature of the surface waters south of Point Conception may be 5°F (2.8°C) higher than that of surface waters north of the Point.

During late winter, spring, and early summer, prevailing north–northwest winds drive the surface waters away from the coast of California and cause cold water to be drawn, or upwelled, from moderate depths to replace it. Upwelling occurs at many locations along the coast but is particularly intense where the coastline extends out as capes and points. A conspicuous center of upwelling exists near Morro Bay, which is about 60 miles north of Point Conception. Here, due to upwelling, the mean temperature of surface waters from March through June can be about 10°F lower than that of surface waters south of Point Conception.

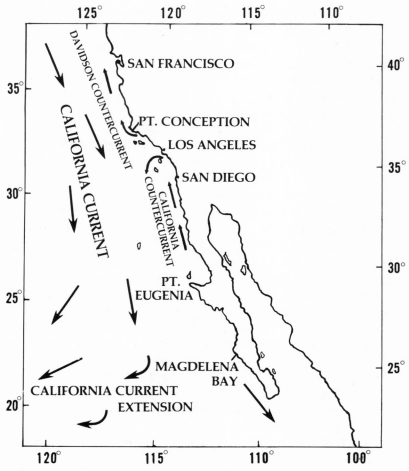

Fig. 14. Oceanic currents along coast of California and Baja California. (After Boyd)

Where the waters of high and low temperature meet and merge, they are readily distinguishable. For upwelling of cold water brings nutrients to the surface, and upon these nutrients plankton feeds. The plankton becomes so plentiful that it gives to the waters a greenish hue. In warmer waters, where there is no upwelling, there is a deficiency of nutrients, plankton does not develop in such numbers, and the waters remain a clear cobalt blue.

Differences in surface termperature of waters to the north and south of Point Conception are reflected in differences in the fauna, since temperature is the physical factor of greatest importance in determining the distribution of marine organisms.

Spiny Lobsters, Shrimps and Crabs of the West Coast

Within the prevailingly warm waters south of Point Conception lives the California spiny lobster, *Panulirus interruptus*, which occupies one of the most northern ranges of its family, the Palinuridae, and is the only species of lobster along the coast of California. The range of this spiny lobster extends from Monterey Bay to the Gulf of Tehuantepec, but the animal is common only as far south as Magdelena Bay, Baja California. Another species, *Panulirus inflatus*, occupies a more southerly range, from Baja California to the Gulf of Tehuantepec. A third species, *Panulirus gracilis*, occurs from the Gulf of California to Peru and the Galápagos Islands.

For *Panulirus interruptus*, Baja California is the center of greatest population density. Yet this spiny lobster is abundant enough along the coast of southern California to support an important fishery. South of Point Conception, most areas in which the bottom is rocky and thickly covered with kelp are intensively fished for this lobster. The fishermen take their catch with traps, or pots, which are made with a wooden frame covered by wire or by wooden laths. As in all spiny lobsters, the meat of the tail is the part used for food.

Because the California Current contains cool subarctic water, many northern species of marine life range far south along the western coast of North America. Among these are nine species of shrimps of the genus *Pandalus* that are found and fished off the west coast of North America as far north as the Bering Sea. The range of three of these species extends southward well into California.

A large, commercially important crab, *Cancer magister*, which often attains nine inches in body width, inhabits relatively shallow, sandy areas in cold water off the west coast of North America. Known in California as the market crab, it is

called the Dungeness crab in more northern regions, after a
small fishing village on the strait of Juan de Fuca, Washington, where commercial fishing for this crab started.

Although Dungeness crabs may be found from the Aleutian Islands southward to Magdalena Bay, Baja California, the

**Fig. 15. Average king crab caught in American fishery is less than
half the size of crab shown here. This is a large one, measuring
about six feet from tip of one long leg to tip of opposite leg.
(Courtesy of Wakefield/Pacific Pearl Seafoods)**

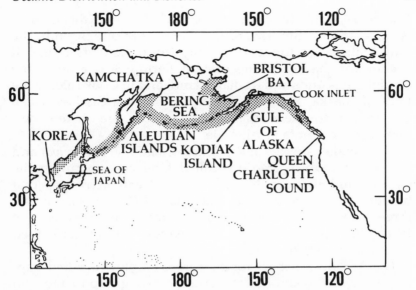

Fig. 16. Distribution of king crab in North Pacific Ocean.

crabs are scarce in warm waters beyond Point Conception. Fishing areas extend southward only to the Monterey Peninsula, about 200 miles north of Point Conception. In southern California, where the Dungeness crab is only occasionally captured, several other species of *Cancer,* all going under the name rock crab, are fished and sold.

Dungeness crabs are trapped largely in baited round wire pots, set in strings of up to 60. In processing, the crabs are boiled and marketed whole, or the cooked meat is removed for direct sale or for canning. A substantial market for live Dungeness crabs also exists.

Crabs of the North Pacific Ocean

Among economically important decapod crustaceans of the North American west coast, one of the most conspicuous is the king crab, *Paralithodes camtschatica*. This species occurs from Korea and the Sea of Japan northeastward to Kamchatka and into the Bering Sea, eastward along the Aleutian chain to

Bristol Bay and the edge of the pack ice, into the Gulf of
Alaska, and southeastward to Queen Charlotte Sound. The
king crab is not found uniformly throughout its range. It
occurs abundantly in the eastern Bering Sea, around the
Shumagin Islands and Kodiak Island, and in Cook Inlet in the
Gulf of Alaska. Here the principal fishing areas have been
situated. In southern Alaska, the crab has been taken commercially only on a small scale.

Two smaller species of *Paralithodes*, namely, *P. brevipes* and
P. platypus, which also are called king crab, occur in the
eastern and western North Pacific Ocean in ranges that
broadly overlap each other and that of *P. camtschatica*. All
three species have been used in the crab canning industry of
Japan, but only *P. camtschatica* is widely sold in the United
States as frozen king crab.

The American fishery for king crabs employs a baited trap,
or pot, which, when full of crabs, may weigh up to a ton. In
the past, American fishermen used two other kinds of gear:
tangle nets, which are eight feet high and 150 feet long and are
often set in parallel rows, five to ten miles long and 1500 yards
apart, to entangle king crabs as they move along the ocean
bottom; and otter trawls, which scoop masses of king crabs
into a huge bag as it is towed along the ocean floor. But in the
American fishery, tangle nets were outlawed in 1955 and otter
trawls in 1960, on the grounds that they are nonselective and
entrap females and young males as well as the large males
that are kept for processing. Japanese and Russian fishermen,
who were not so restricted, continued to use both types of
gear. In recent years, however, bilateral agreements between
the United States and Japan and between the United States
and the Soviet Union have greatly reduced the use of tangle
nets and trawls in the eastern Bering Sea and in the Gulf of
Alaska.

The average king crab of the American fishery weighs
about 10 pounds, but occasionally a catch may include a crab
weighing 25 pounds. Among the largest king crabs ever
caught was one weighing 28 pounds and measuring six feet
two inches from the tip of one leg to the tip of the opposite leg.
The meat from "shoulders," legs, and claws is removed after

Fig. 17. Fishing for king crabs in Aleutian Islands near Adak. A king crab pot, such as this, often holds about 100 crabs weighing as much as seven or eight pounds each. (Bob and Ira Spring)

cooking and processed for food. A king crab that weighs 10 pounds when alive yields about two pounds of cooked meat.

It may seem strange that king crabs, which require very cold water, extend southward to the coast of Korea (about 35°N) in the western Pacific but are limited to the area around Queen Charlotte Sound (about 50°N) in the eastern Pacific. The reason is related to the nature of oceanic currents that exist in these regions.

In the western Pacific, there is a current of cold subarctic water that is known as the Oyashio, a term that means "parent current." The Oyashio flows out of the Bering Sea and continues southward close to the northeastern coast of Japan until it reaches approximately 35° north latitude. Here the Oyashio terminates by mixing with the warm Kuroshio, or Japan Current, which flows northeastward from Taiwan and the Ryukyu Islands. The term Kuroshio means "black stream," an allusion to the deep azure blue of its waters.

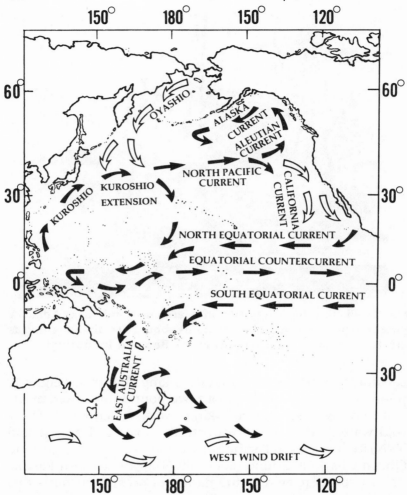

Fig. 18. Major currents of the North Pacific Ocean and the central and western South Pacific. Black arrows: warm currents; white arrows: cool or cold currents.

Off Japan the Kuroshio turns eastward as the Kuroshio Extension. Then, in the eastern Pacific, a continuation of the Kuroshio Extension becomes part of the North Pacific Current, which divides into a branch that flows south as the California Current and one that flows north into the Gulf of Alaska as the Aleutian Current.

The North Pacific Current contains both warm Kuroshio water and cold subarctic water; hence it has a moderating influence on northern and southern regions alike. The portion that flows north into the Gulf of Alaska as the Aleutian Current moderates the climate of coastal southern Alaska. In temperature characteristics, therefore, a latitude of 35°N in the western Pacific and one of 50°N in the eastern Pacific are not so very different. This explains why the southern limit for the king crab *Paralithodes camtschatica* occurs at these very different latitudes on the two sides of the Pacific Ocean.

Another economically important crustacean of the Bering Sea and of the northwest coast of North America as far south as British Columbia is the tanner crab, consisting of two species, *Chionoecetes bairdii* and *C. opilio*. The tanner crab is a spider crab. Particularly, *Chionoecetes bairdii* is taken by Japanese and Russian fishermen in the eastern Bering Sea and by American fishermen both in the eastern Bering Sea and in the area around Kodiak Island, Alaska. Smaller numbers of *C. opilio* appear in the catch. To the south, mainly off Washington and Oregon, the species *Chionoecetes tanneri*, which also is known as the tanner crab, is commonly taken.

The American fishery for tanner crab remained incidental to the king crab fishery until 1968, when reduced catches of king crabs and restrictions upon the fishery prompted many Alaskans to fish for tanner crabs with the use of specially designed traps.

In the Sea of Japan there has long been a large and valuable fishery for *Chionoecetes opilio elongatus,* known to the Japanese as zuwai-gani ("gani" means "crab") and, more recently, for *C. japonicus,* known as benizuwai-gani. Much of the Japanese catch from the Sea of Japan and the Bering Sea is canned and sold as snow crab in the United States and as zuwai crab in the United Kingdom.

Chionoectes opilio (but not the subspecies *elongatus*) is found both in the Gulf of Alaska and the Bering Sea and also off the northeast coast of North America from Greenland to the Gulf of Maine. In the North Atlantic, *C. opilio* is usually known as the snow crab or queen crab. Starting in 1967, this crab has become the basis of a rapidly growing Canadian fishery that is centered largely on the east and southeast coasts of Newfoundland.

The Gulf Stream System

Formerly, it was thought that the Gulf Stream was a discharge stream for water accumulating in the Gulf of Mexico, which was presumed to be a basin-like area. In the view of modern oceanographers, however, the Gulf Stream is part of a discharge of water masses that accumulate on the western side of the Atlantic because of the tangential stress of the northeast trade winds. The Kuroshio, the Brazil Current, the East Australia Current, and the Agulhas Current off the southeast coast of Africa are discharges of similar origin.

Over much of the Atlantic Ocean that lies between 35° and 15° north latitude, the trade winds blow steadily from the northeast toward the southwest, particularly in winter. This region is ideal for sailing, there is little rain, few clouds float in the sky, the breeze is fresh, the ocean is deep-blue in color, and the swells are long and are tipped with bright crests of foam.

The steady northeast to southwest flow of air drives ocean water before it, so that the water piles up at the western edge of the North Atlantic Ocean—which is, of course, the eastern edge of the North American continent. Here as the Florida Current, the water discharges northward along the continental slope to Cape Hatteras, then, as the Gulf Stream, northeastward to the Grand Banks off Newfoundland, and finally, as the North Atlantic Current, eastward across the North Atlantic to Europe, where it penetrates this land mass from the North Sea to the Mediterranean. These are the principal currents that constitute the Gulf Stream System.

To the right of the Florida Current as it sweeps northward lies the Sargasso Sea, a mass of warm water extending downward about 2250 feet and slowly drifting to the southwest under the influence of the northeast trades. It is partly water of the Sargasso Sea that is returned to the north and east in the discharge of the Gulf Stream system.

The Florida Current acts as the western boundary of the

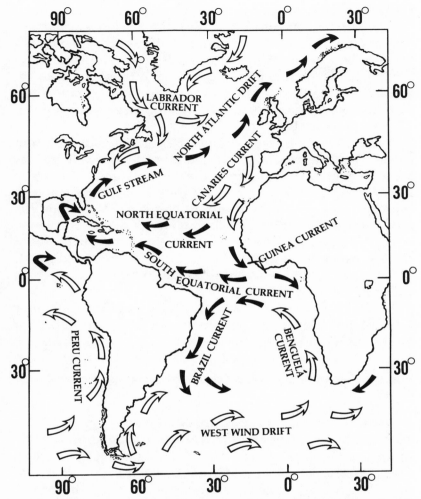

Fig. 19. Major currents of the Atlantic Ocean and the eastern South Pacific. Black arrows: warm currents; white arrows: cool or cold currents.

Sargasso Sea. Off the Grand Banks, where the Gulf Stream swings first to the northeast and then more to the east, its intensity of flow acts as a barrier that prevents the warm water of the Sargasso Sea from overflowing the cold water of the North Atlantic Ocean. As the American oceanographer

Henry Stommel has pointed out in his book *The Gulf Stream*, contrary to popular conception, the Gulf Stream is not a river of hot water. Indeed, its temperature differs little from that of the Sargasso Sea to the right of its direction of motion. According to Dr. Stommel, the role played by the Gulf Stream in moderating the climate of western Europe lies primarily in its capacity to determine the northern boundary of warm water in the Sargasso Sea.

To the north of Cape Hatteras, the cold Labrador Current sweeps southward along the continental slope and carries polar and subpolar water from the coast of Labrador to Cape Hatteras. This slope water, as it is often called, lies between the Gulf Stream and the shores of North America from Virginia northeastward. Cold slope water is rich in nutrients, rich in plankton, and a deep green in color. Where it meets the Gulf Stream in a boundary that is known as the "cold wall," green slope water contrasts strikingly with the cobalt–blue water of the Gulf Stream.

As we shall now see, the Gulf Stream and the Labrador Current play key roles in determining the distribution of marine fauna and flora along the east coast of North America.

Shrimps and Spiny Lobsters of the North American East and Gulf Coasts

Many species of shrimps are found and fished off the shores of North America, but in commercial importance none rivals three species of penaeids occurring along the coast of the South Atlantic states and in the Gulf of Mexico. These are the white shrimp *Penaeus (Litopenaeus) setiferus*, the brown shrimp *Penaeus (Melicertus) aztecus aztecus*, and the pink shrimp *Penaeus (Melicertus) duorarum duorarum*.

These three species, along with several other species of penaeids that are not nearly so important commercially, characteristically inhabit warm southern waters. Fisheries for these shrimps exist on the Atlantic coast only as far north as Beaufort, North Carolina. This occurs because north of Cape Hatteras, where the Gulf Stream swings offshore to the northeast, cold slope water replaces warm subtropical water along the shore. In the cold slope water, the numbers of southern shrimps decline so greatly that fisheries for them are not profitable.

Nevertheless, occasionally in late summer or early autumn there are reports of large numbers of penaeid shrimps in northern waters. An instance of such an influx occurred in 1957. On October 19 of that year, Mrs. Francis H. Low reported to the American Museum of Natural History that many shrimps were being caught in Shinnecock Bay, Long Island, in traps that had been set for eel, fluke, and flounder. Several thousand pounds of shrimps had already been taken during September and the first half of October. Sample specimens from the catch were later identified by Dr. Fenner A. Chace, Jr., as the brown shrimp *Penaeus (Melicertus) aztecus aztecus*. Apparently, local influxes may occur occasionally as far north as Woods Hole, Massachusetts, but seldom are the shrimps present in such numbers.

In areas where southern penaeid shrimps are plentiful, they support a fishery that ranks as one of the most important in the United States, exceeding any other fishery in value. During some years the catch may represent over 90% of the total shrimps landed in the United States.

The southern shrimp fishery includes both a commercial bait fishery for juvenile penaeid shrimps of the inshore waters and the commercial fishery proper, mainly for large mature penaeids of the coastal offshore waters. The latter fishery is the core of the shrimp industry. With the exception of some waste that is ground into meal, the entire catch of the offshore fishery is processed for food. Otter trawls make the catch. Tails of large shrimps are marketed fresh or frozen, while those of smaller shrimps are dried and canned.

In recent years, large concentrations of royal-red shrimps, *Hymenopenaeus robustus,* have been found in three areas along the Continental Slope of the western Atlantic: east of St. Augustine, Florida; south–southwest of the Dry Tortugas; and southeast of the Mississippi River Delta. Fishermen have taken royal-red shrimps from these areas and have marketed

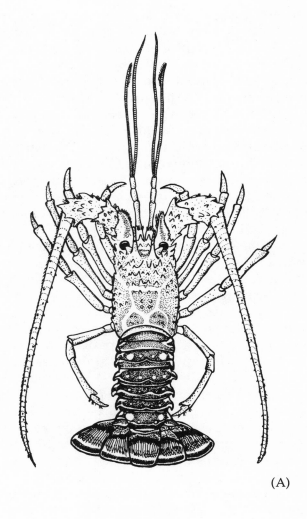

(A)

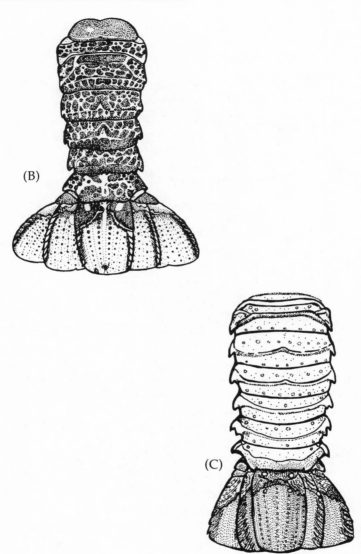

Fig. 20. Four kinds of lobster tails generally available in markets. (A) Florida spiny lobster; (B) South African rock lobster; (C) Western Australian rock lobster; (D) California spiny lobster. Entire lobster is shown in (A), only tails in (B), (C), (D). [(A), after Manning]

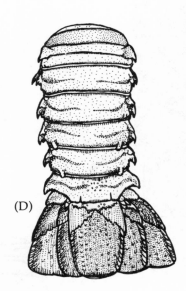

(D)

them fresh-frozen or iced. These shrimps occur from North Carolina to the Guianas. They live on mud, sand, or silty sand at depths of 900 to 1800 feet.

Another commercially important species of decapod crustacean, the Florida, or Caribbean, spiny lobster, *Panulirus argus,* is found in warm waters along the Atlantic and Gulf coasts of the Americas as far south as Rio de Janeiro, Brazil. On the east coast of southern Florida, off the Florida Keys, in the Bahamas, and off Belize, sizable numbers of this spiny lobster are caught commercially in wooden or wire traps. Elsewhere in its range, including Bermuda and on the west coast of Florida, this species is primarily of local importance as an item of food.

For *Panulirus argus,* the northeast swing of the Gulf Stream away from the shore and the replacement of warm subtropical water by the cold slope water of the Labrador Current effectively establishes a northern limit, which lies close to Beaufort, North Carolina.

Faunal Change Points

For a number of marine animals that are found along the southeastern coast of North America, Cape Hatteras does not mark the northern limit of distribution. On the contrary, many of these animals range northward to Cape Cod and some occur as far north as Newfoundland.

Three geographical localities, Cape Hatteras, Cape Cod and Newfoundland, are so-called change points for marine flora and fauna of the North Atlantic coast. At each of these localities, the isotherms, that is, the lines that are found on certain maps to denote regions of equal temperature, are crowded closely together, indicating a rapid change in temperature within a short distance over the sea.

In 1881, the well-known botanist Dr. W. G. Farlow, with reference to the marine algae of New England and nearby coasts, pointed out that north of Cape Cod the marine vegetation has an arctic quality and is a continuation of the flora of Greenland and Newfoundland. South of Cape Cod, Dr. Farlow continued, arctic forms have disappeared and algae characteristic of warmer seas are present. He added that the differences between the flora of Massachusetts Bay, which borders the north shore of Cape Cod, and that of nearby Buzzards Bay, which juts well up into the south shore of Cape Cod, are greater than the differences between the flora of Massachusetts Bay and that of the Bay of Fundy in Nova Scotia or between the flora of Nantucket and that of Norfolk, Virginia.

Why does a change point for fauna and flora of the Atlantic coast exist at Cape Cod? The geography of oceanic currents provides a clue. Once the Labrador Current has swept southward past the southeastern coast of Nova Scotia, its cold arctic water, essentially unmixed with warm water of the Gulf Stream, enters the Gulf of Maine, which is bounded by southwestern Nova Scotia and the bay of Fundy on the north and Cape Cod on the south. Since the water of this area is cold, the flora and fauna are arctic and subarctic in character.

Slope water carried by the Labrador Current past Cape Cod undergoes a gradual mixing with warm water that is carried north by the Gulf Stream and eddies shoreward. Such mixing raises the temperature of the slope water. Thus, the water that bathes the south shore of Cape Cod, the shores of Rhode Island, and those of Connecticut, Long Island, and regions to the south is warm compared with water of areas to the north.

Blue Crabs, Stone Crabs and Green Crabs

An animal that has crossed the first change point at Cape Hatteras with ease and the second at Cape Cod with difficulty is the blue crab, *Callinectes sapidus*. The range of the blue crab extends to Uruguay, but south of Texas this species is less common than are some other species of the genus *Callinectes*. Sizable fisheries for blue crabs exist in the South Atlantic and Gulf states. Blue crabs are most numerous, however, in and around Chesapeake Bay, and it is here that the multimillion dollar blue crab fishery is centered.

Formerly, as far north as Cape Cod, blue crabs were plentiful enough to support minor fisheries. But in recent years the crabs have declined greatly in number. Now, in Narragansett Bay, for instance, catches of blue crabs are insignificant.

North of Cape Cod, blue crabs are even more scarce and are without commercial value. Yet from 1948 to 1956 many blue crabs were captured as far north as Casco Bay and the mouth of the Kennebec River in Maine. During those years, according to officials at the United States Bureau of Commercial Fisheries, surface temperatures of coastal waters in Maine were higher than normal, making it possible, no doubt, for migrant crabs from south of Cape Cod to survive in normally cold waters off southern Maine.

Blue crabs may also be taken in some parts of Europe. During the years 1900 to 1968, incidental finds were reported from France, The Netherlands, Denmark, Germany and Italy. By 1969, the crabs were firmly established in Greece, Turkey, Lebanon, Israel and Egypt. Subsequently, it was reported that millions of blue crabs lived in lakes of the Nile Delta, where they consumed fish and often damaged fishing nets.

In the United States, catches of blue crabs are made with crab nets, pots, trotlines, scrapes, and other gear. The winter fishery for blue crabs depends on the crab dredge, which digs

Fig. 21. Floats containing blue crabs about to shed their shells. Electric lights strung above floats permit night workers to remove crabs at the critical time in the shedding stage. Crabs are then packed in ice, on docks in background, for shipment to market. (Courtesy of the Virginia Institute of Marine Science, Gloucester Point, Virginia)

up crabs that have burrowed into the mud in deep water. Peeler crabs, that is, crabs about to shed their shell, are collected in the warmer months of the year. These crabs are held in shedding floats until an hour or so after ecdysis, when they are packed in ice and seaweed and shipped alive in refrigerated cars or trucks, to be marketed as soft-shelled crabs. Hard-shelled crabs are marketed alive and iced, or the cooked meat is removed and sold in iced containers.

Another crab that is found along the South Atlantic and Gulf coasts, from North Carolina to Yucatan, Cuba, Jamaica and the Bahamas, is the stone crab, *Menippe mercenaria*. This large, sluggish, colorful crab lives around jetties, where it wedges itself into the sand at the edge of rocks; this crab is also found burrowing just below the tide mark in shallow areas.

Fig. 22. Soft blue crabs being removed from shedding float. Because shell of crab becomes papery within four to six hours following shedding, floats must be examined frequently. (Courtesy of the Virginia Institute of Marine Science, Gloucester Point, Virginia)

The crab is caught mainly in traps, or pots. Since there is little meat in the body, fishermen usually break off one or both of the powerful claws and return the live crab to the water. This enables the crab to regenerate its claws for subsequent harvesting. The meat of the claws is marketed either boiled and chilled or fresh-frozen. It is considered a great delicacy. Although stone crabs are occasionally caught and the meat offered for sale in various South Atlantic and Gulf states, the commercial fishery is restricted almost entirely to Florida.

Like the blue crab, the green crab, *Carcinus maenas,* presents an interesting case of a marine animal that gradually extended its range northward. In 1874, Sidney I. Smith, an authority on crustaceans of the Atlantic coast, gave the range of the green crab as New Jersey to Cape Cod, adding that no other common crustacean along the Atlantic coast had, to his knowledge, such a restricted range.

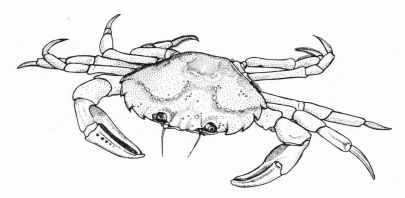

Fig. 23. Green crab, relentless enemy of tasty soft-shell clam, is used by sports fishermen as bait.

Yet by 1905 green crabs were reported in and near Casco Bay, Maine. Twenty-five years later, when the distinguished American taxonomist Dr. Mary J. Rathbun published a classic monograph on cancroid crabs of America, she stated that green crabs were abundant even farther to the northeast—in George's River, Thomaston, at the entrance to Penobscot Bay.

The year 1937 found the green crab established at Seal Cove on Mount Desert Island and two years later in Frenchman Bay. By 1951 the crab could be collected in Passamoquoddy Bay near Digdeguash, which is just over the Canadian border in New Brunswick. Shortly thereafter, it was discovered near Halifax, Nova Scotia.

Why was the green crab able to extend its range northeastward beyond Cape Cod? Possibly larval crabs were carried by local currents, or adult crabs traveled along the shore, or lobster and sardine fishermen inadvertently transported the crabs in boats and trucks. Yet the green crab would never have become established north of Cape Cod if environmental conditions had been unfavorable for survival. As mentioned earlier, surface temperatures in coastal waters off Maine and vicinity during the late 1940s and early 1950s were higher than normal. This may have been responsible for the occurrence of the green crab so far north at that time.

Subsequent to the early 1950s, this crab was not reported in

Canadian waters, and its numbers in the waters of Maine declined. According to officials at the United States Bureau of Commercial Fisheries, the decline may have been due to the much lower temperatures of those waters in the later years.

The green crab has considerable economic importance in a negative sense. It burrows into tidal flats of sand and mud and, with its large claws, seizes and crushes its prey, which includes large numbers of soft-shelled clams *Mya arenaria* and quahogs *Mercenaria mercenaria*. These species of mollusks are favored as seafood delicacies. On the positive side, green crabs are useful as bait.

Northern Shrimps and Crabs of the North American East Coast

Off coastal areas of northeastern North America that are swept by the Labrador Current lives *Pandalus borealis*, the northern shrimp, which, because of its color, is also known as the pink shrimp. This species is circumpolar, living in the depths of boreal and subarctic seas throughout vast areas of the Eastern and Western Hemispheres. Commercial fisheries for this shrimp exist in Alaska, British Columbia, Norway, Sweden, British Isles, Iceland, southern Greenland, Newfoundland, the Canadian Maritimes, and northern New England.

For several decades, the pink shrimp was plentiful enough off the coast of Maine to support a small but profitable winter

Fig. 24. Northern shrimp, also known as pink shrimp, occurs abundantly in boreal and subarctic seas and is the basis for extensive commercial fisheries. (After Scattergood)

fishery. The catch, primarily taken by otter trawls, mainly consisted of large egg-bearing females, which congregate near the mouth of estuaries and on other inshore grounds prior to the time when they hatch their eggs. In the waters off Maine, hatching of eggs reaches a peak in March and April.

An extraordinary feature of the Maine shrimp fishery— and of a similar fishery off Gloucester, Massachusetts— was its rise in the late 1930s, its peak period of production in 1944 and 1945, its rapid demise during the late 1940s and early 1950s, its revival to another peak in 1969, and again its decline. In recent years, efforts have been made to regulate mesh size of nets and length of the open season for fishing in order to help sustain this fishery.

Another fishery—a minor one for bait shrimp—has existed along the coast of New York and New Jersey. Several species of small, transparent "grass" or "glass" shrimps, *Palaemonetes,* and also the sand shrimp, *Crangon septemspinosa,* have been caught.

On the coast of Maine and Massachusetts there has been a small fishery for the Jonah crab, *Cancer borealis,* which occurs from Nova Scotia to Florida, and for a rock crab, *Cancer irroratus,* found from Labrador to South Carolina. This crab fishery has been largely a by-product of the lobster fishery, most crabs being caught in traps set for American lobsters.

Parenthetically, it may be noted here that fishermen in the

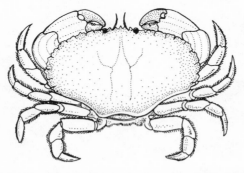

Fig. 25. Jonah crab, found along east coast of North America and taken largely as a by-product of fishery for American lobsters. (After A. B. Williams)

British Isles also take crabs in traps used for lobsters. Known
locally as creels, these traps may be laid in shallow water for
lobster and at depths of over 60 feet for the crab *Cancer
pagurus*. This species occurs from the northwest coast of Norway south to Portugal. French fishermen may take this same
species of crab in nets baited for spiny lobsters.

**Fig. 26. Lobster fisherman on coast of Maine prepares to set his
traps. (Courtesy of the Maine Department of Marine Resources)**

True Lobsters and Red Crabs

From southeastern Labrador to southern New England (but particularly in waters of Maine, southern Nova Scotia, and the southern Gulf of St. Lawrence) the American lobster, *Homarus americanus*, supports an important inshore fishery. Lobsters taken in these areas generally weigh about a pound or two and are caught in traps, or pots, that are set from low tide to a depth of about 120 feet.

An inshore fishery for American lobsters has existed for a very long time, with coastal Indians and, subsequently, French and English explorers of the early seventeenth century taking American lobsters for use as food. The settlers of Massachusetts Bay also ate lobsters, noting in their records the large size and great numbers of these animals.

Today the inshore fishery for American lobsters is operating at maximum capacity, but is yielding less and less. Increasing consumption of lobsters, coupled with overfishing and general pollution, is steadily pushing demand past supply and is resulting in a rapid escalation of the price of plate-sized lobsters ("selects," weighing 1½ to 3 pounds). These are now luxury items.

In recent years sizable concentrations of large American lobsters have been taken in submarine canyons along the edge of the American continental shelf from the southeastern part of George's Bank, off Massachusetts, southward to the latitude of Cape Hatteras, North Carolina. These lobsters, occurring at depths of about 300 to 2400 feet, are the basis of what was formerly a trawl fishery and is now mainly a pot fishery of steadily increasing value. The pots are set in strings of 25 to 75, spaced 60 to 180 feet apart, and covering two to three miles.

Many lobsters caught in these pots weigh 10 to 15 pounds, some 25 pounds or more. These giants are thought to be about 15 to 20 years old, although some estimates place the largest as 40 to 50 years of age. In 1956, a 44½-pound American lobster reportedly was hauled from deep water off the eastern end of Long Island, New York. The *authenticated* record for

Fig. 27. Large American lobsters, weighing 10–15 pounds, taken in pots set in deep water near edge of continental shelf. (© National Geographic Society)

size, however, appears to be held by an American lobster over three feet in length that, when alive, weighed 42 pounds 7 ounces. This lobster, captured in 1934 off the Virginia Capes, is on display at the Museum of Science in Boston, Massachusetts.

For those readers who wish to learn more about the dimensions of very large lobsters, I recommend a short article by Dr. Torben Wolff entitled "Maximum size of lobsters *(Homarus)* (Decapoda, Nephropidea)," which was published in 1978 in the journal *Crustaceana* (vol. 34, pp. 1–14). This article contains many data on the length and weight of giant lobsters and the size of their big claws.

A valuable by-product of the offshore fishery for American lobsters is the deep-sea red crab, *Geryon quinquedens*, which occurs in large numbers on the edge of the continental shelf from Nova Scotia to Cuba. With the use of lobster pots set in strings of about 50, deep-sea red crabs can most readily be taken in water of 1200 to 2100 feet, where the temperature ranges from 38 to 43°F (3.3–6.1°C). Until recently, there was no commercial fishery, since deep-sea red crabs are fragile and, when not chilled, are difficult to keep alive during a long run home from the fishing grounds. Now, however, new onboard refrigeration systems can reduce loss of crabs during transit. As a result, a processing plant for deep-sea red crabs is in operation in Gloucester, Massachusetts. The picked meat, which is sold in a canned or frozen state, is now available throughout the United States as a delicious and less expensive alternative to lobster. In winter and occasionally at other times of the year, deep-sea red crabs can also be purchased alive at or near the port at which they are landed.

The European lobster, *Homarus gammarus*, which is a near relative of the American lobster and closely resembles it, occurs off the northwestern coast of Europe. Fisheries for this lobster exist along rocky shores in southern Norway and the British Isles, in isolated rocky areas off the shores of Denmark and Germany, and in rocks at the foot of dikes in The Netherlands. The lobster may be trapped as far south as Portugal, and it is present but not plentiful in the Mediterranean Sea.

Another fishery along the west coast of Europe is that for the Norway lobster, *Nephrops norvegicus*, which burrows into soft, sticky mud at water depths of over 1000 feet and is

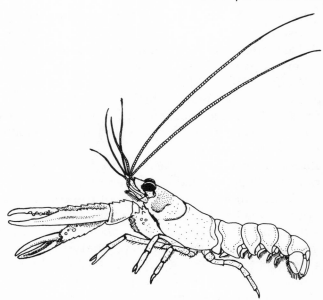

Fig. 28. Norway lobster, dweller of deep waters along west coast of Europe, is a popular item of food.

caught mainly with otter trawls. Brilliantly colored in yellow and orange, this lobster can be distinguished from the American and European lobsters by its claws, which are long and slender and are armed with several rows of strong tooth-like spines. The Norway lobster is less bulky than the other two, its slender shape being responsible for one of its popular names, the maiden lobster. This animal is also known as the Dublin Bay prawn.

Fisheries for the Norway lobster exist throughout its range, which extends from northern Norway into the Mediterranean. Yet the fisheries are concentrated in the Bay of Biscay, in the North Sea, and off the western coast of the British Isles, where the Norway lobster is most abundant. The catch is generally marketed whole, but frequently in Great Britain only the tails are sold, mainly shelled and frozen. As a luxury product, this lobster is often sold under the name scampi, after an Italian dish of the same name in which it is a principal ingredient.

In 1962, *Nephropsis aculeata,* a close relative of the Norway lobster, was introduced into markets along the east coast of Florida as the "Danish lobsterette." Now called the Florida lobsterette, this species, along with *Metanephrops binghami,* the Caribbean lobsterette, is abundant in certain areas of the southwestern North Atlantic, Gulf of Mexico, and Caribbean Sea; and it appears intermittently in the markets. The large quantities of lobsterettes in these areas should justify the establishment of a commercial fishery, preferably one supplemental to that for royal-red shrimps, which are taken in the same general areas.

Spiny Lobsters and Slipper Lobsters

Widely distributed along the coast of western Europe are two species of lobsters that belong to the family Palinuridae and are known familiarly as spiny lobsters, rock lobsters, and marine crayfishes. One species, *Palinurus elephas,* occurs from the Orkney Islands and the Hebrides off northern Scotland to Spanish Sahara and in the Mediterranean.

Although not exported from Europe to any great extent, *Palinurus elephas* serves as a basis of a more or less important fishery throughout its range. The animal is known by a variety of popular names, including *langouste* in continental Europe and kreeft in The Netherlands. The French term *langoustine,* however, refers to the Norway lobster, *Nephrops norvegicus.*

A second species of spiny lobster, *Panulirus regius,* is found in the western Mediterranean and south along the west coast of Africa to Angola. This lobster, which is often marketed in France, is known popularly as the *langouste royale.*

Numerous other species of spiny lobsters occur in various parts of the world. Except for *Palinurus elephas,* referred to above, and the genus *Jasus,* which lives in temperate regions of the southern hemisphere, all inhabit tropical or subtropical seas. A continuous band drawn around the globe between 40°N and 20°S latitude would encompass most species of spiny lobsters. The lobsters are not present in all geographical areas within the band, but they probably do occur along most

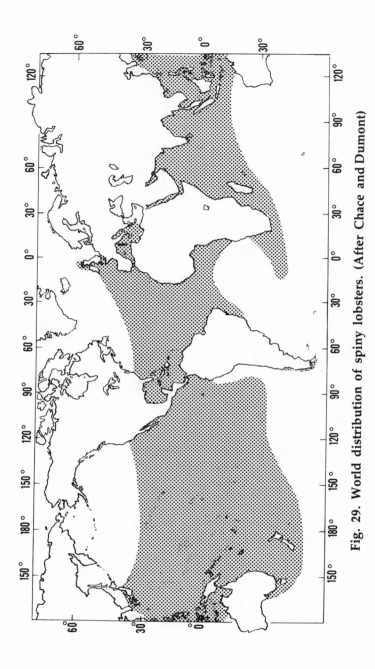

Fig. 29. World distribution of spiny lobsters. (After Chace and Dumont)

coasts that provide adequate food and rocks or reefs within which the spiny lobsters can hide.

Among the spiny lobsters well known to Americans are *Panulirus interruptus,* which is fished commercially off southern California, and *Panulirus argus,* which is taken off Florida and in the Bahamas. Both species have been discussed earlier in this chapter. Commonly available in American supermarkets as frozen "craytails" is *Panulirus longipes cygnus,* which occurs only off Western Australia and is the basis of one of the most important Australian fisheries.

Marketed even more widely in the United States are frozen tails of the South African rock lobster, *Jasus lalandii.* This spiny lobster is found in areas washed by the Benguella Current, which flows slowly northward along the west coast of southern Africa and is particularly well developed from the southern tip of Africa to latitude 18°S; that is, to approximately the border of South-West Africa (Namibia) and Angola. Because prevailing winds blow surface water away from the coast and upwelling takes place, the Benguella Current is a more or less continuous band of cold water that parallels the coast and extends outward for about 120 miles. It is rich in nutrients, plankton, and animal life and supports some of the largest and most prosperous fisheries in the world.

Although the South African rock lobster occurs as far east as Algoa Bay, it is fished commercially only to the west of Cape Point, which lies at the tip of the Cape of Good Hope. Areas washed by the warm Agulhas Current east of Cape Point may be virtually devoid of rock lobsters. Commercial fishing extends northward to Cape Cross in South-West Africa.

South African rock lobsters usually are found between low-water mark and depths of 120 to 150 feet. They appear to seek out areas where the bottom is rocky and is covered thickly with kelp. Rock lobsters have a strong shell, so they readily withstand buffeting by moderate waves, but heavy swells cause them to seek shelter in dense seaweed. Fishermen, working from a dinghy, make the catch with a baited hoop net, which rests upon the bottom and is raised at intervals so that the rock lobsters can be removed.

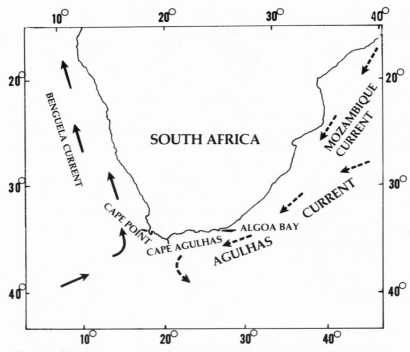

Fig. 39. Oceanic currents along coast of South Africa. Rock lobsters are found as far east as Algoa Bay, but are fished commercially only to the west of Cape Point in areas washed by cold Benguella Current. Solid arrows: cold currents; broken arrows: warm currents.

Closely related to spiny lobsters are the slipper lobsters. These rather strange-looking creatures go by a variety of common names, most of which are derived from the flat, squarish body and the short, flat, broad antennae that make the animal look like a slipper, a shovel, or even a type of bulldozer. Slipper lobsters live in the ocean on level bottom of mud, sand, or rock and sometimes also in reef areas, from the coastline to a depth of 1400 to 1500 feet.

The edible portion of a slipper lobster, as of a spiny lobster, is the muscular tail, which in a slipper lobster is broad and flattened. Reputedly sweeter and tastier than a spiny lobster, slipper lobsters are fished commercially in many parts of the western Pacific and Indian Oceans. Imported "slipper tails" reportedly may sometimes be found in markets and restaurants in the United States.

Although 10 species of slipper lobsters occur in the Western Central Atlantic, there is no established fishery for them. Some large species, particularly of the genus *Scyllarides*, are fished and sold locally or are caught incidental to the fishery for spiny lobsters. One deep-water species, *Scyllarides aequinoctialis*, may appear in small quantities in markets of Puerto Rico and The Netherlands Antilles. Here and elsewhere, this animal may occasionally be found on the dinner table as a gourmet dish.

Crabs and Shrimps of the Indo-West Pacific

The Indo-West Pacific, known more accurately as the Indo-West Pacific Faunal Area, or the Tropical Western Pacific and Indian Ocean Faunal Area, is the world's largest marine shore-lined faunal area. It extends eastward from the

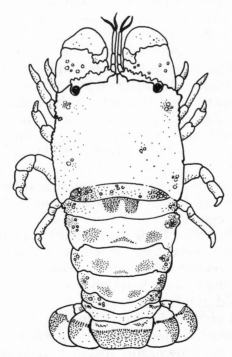

Fig. 31. Spanish slipper lobster, *Scyllarides aequinoctialis*, close relative of spiny lobsters. Slipper lobsters are fished commercially in Indo-West Pacific, and imported "slipper tails" appear occasionally in American markets and restaurants. (After Manning)

coast of Africa and the Red Sea across the Indian Ocean, through the tropical western Pacific Ocean to Polynesia, and northward to Hawaii. Some species of animals occupy this entire area, others a large portion of it, and still others only a restricted region within it.

In the Indo-West Pacific Faunal Area, many species of marine crabs are sold as food, mainly in markets not far from where the crabs are captured. Among these is the red frog crab, or spanner crab, *Ranina ranina*, an unusual looking creature with a broad but tapered body and flattened legs that are well adapted for burrowing backward into sand, where the crab hides. The frog crab occurs from Africa to Hawaii (where it is known as the Kona crab) and is found also in Japan and Taiwan. The crab is taken in nets at depths of 30 to 150 feet and is eaten everywhere in the Indo-Pacific where it may be captured. Its flesh is said to be very tasty.

Less remarkable in appearance, but more easily procured for food, is *Scylla serrata*, the Samoan, or mud, crab, which is also known as the serrate swimming crab. The latter name comes from the 21 notches along the anterior edge of the crab's shell. This large, active animal lives in brackish water on muddy bottom and is captured in crab pots and hoop nets. It is widely distributed throughout the Indian and Pacific Oceans, from Africa to Japan and China, and is commonly available in local markets. Some years ago, the serrate swimming crab was successfully introduced into Hawaii, where an open ecological niche appears to have been available, and the crab now supports a growing commercial fishery.

A most unusual Indo-West Pacific crab is *Podophthalmus vigil*, the long-eyed swimming crab. Its eyes are on such long stalks that, when lying horizontally within a groove along the front margin of the body, they extend almost across the body. This crab lives on soft bottom in estuaries and at the mouth of rivers, where the water is brackish. It is caught in nets and is commonly sold in local markets.

The largest crab in the world, the giant Japanese spider crab, *Macrocheira kaempferi*, lives in deep water off the coast of Japan. When full-grown, this species may be over a foot in body width and its claws may have a total spread of over 10 feet.

Eight species of crabs that inhabit coral reefs in remote but widespread areas of the Indo-West Pacific have recently been found to be highly toxic. The flesh of these crabs, when cooked and eaten, sometimes causes severe illness and even death. All of these species of crabs are members of the family Xanthidae. Four additional species of xanthid crabs and a fifth species of crab of the family Parthenopidae are known from toxicological tests to be mildly toxic. It is not known whether the crabs are poisonous because they eat toxic food or because they generate a poison during their own metabolism. Although any of these species of crabs may appear for sale locally near their place of origin, none is generally marketed.

Within the Indo-West Pacific Faunal Area, about one-half of the world's catch of shrimps is taken. Fishermen often use primitive gear and dugout canoes and market their catch nearby, frequently within walking distance of the village in which they live. In some areas fishermen use beam trawls, with portions of this catch being marketed abroad. Much shrimp fishing is done in brackish water, primarily in lagoons, estuaries, and tributaries within the great delta areas.

For the most part, marine shrimps caught in Asia are various species of the genera *Penaeus* and *Metapenaeus*. One species, *Penaeus japonicus*, which is known as kuruma-ebi, or wheel-shrimp, is among the most esteemed and expensive of shrimps eaten in Japan. There it is used in the preparation of the traditional Japanese dish, tempura. In Tokyo alone, several hundred restaurants specialize in tempura, and many more include tempura on the menu. To meet a large popular demand, Japan not only imports great quantities of shrimps from other countries, but also supports shrimp farming, a subject that is covered in the final chapter of this book.

Australia has well-developed shrimp fisheries for various species of *Penaeus* that occur in estuaries and offshore to the edge of the continental shelf. Since 1947, a steadily increasing fleet of otter trawlers has been operating in areas where large concentrations of shrimps are found, primarily off New South Wales, Western Australia, Queensland, and Northern Territory. Sizable quantities of frozen shrimps are exported, particularly to Japan and to a less extent to the United States and South Africa.

South American Shrimps

In warm waters off continental South America, ten species of shrimps belonging to the genus *Penaeus* account for about nine-tenths of the total South American catch. Commercial fisheries for these penaeid shrimps extend along the eastern coast from the Gulf of Mexico to southern Brazil and along the western coast from Baja California to northern Peru. The southern limit of the range of these penaeids appears to be Uruguay on the east coast and Peru on the west coast.

Reference to a map reveals that these penaeid shrimps extend about 20° of latitude farther south on the east coast of South America than they do on the west coast. The reason is that a warm, southward-flowing current, the Brazil Current, parallels the shore along the east coast, enabling the penaeid shrimps that are tropical and subtropical in character to range south to Uruguay. Along the west coast, a cold current, known as the Peru, or Humboldt, Current, swings northward past the shores of Chile and Peru and effectively limits the distribution of these penaeid shrimps to northern regions. In some areas, the chill of the Peru Current is enhanced by additional cold water brought to surface layers through upwelling.

In cold water of the South Atlantic off southern Brazil and Argentina there are two species of penaeid shrimps, the small red *Artemesia longinaris* and the large red *Hymenopenaeus muelleri*. These shrimps occur in large numbers along the Argentinian coast south of latitude 35°S. They are taken by beam and otter trawlers from waters of up to 60 feet and are largely marketed and consumed locally.

In cold water of the South Pacific off Chile, three species of shrimps, two of which belong to the family Pandalidae, are taken commercially by otter trawlers, with the entire catch

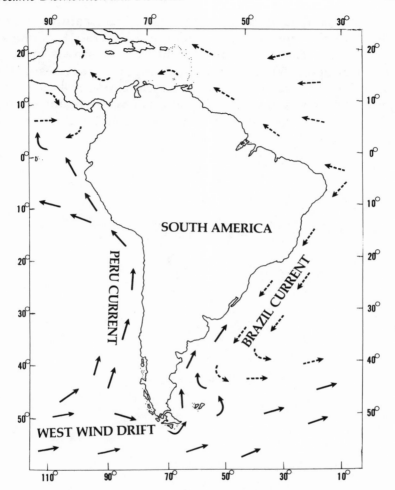

Fig. 32. Oceanic currents along the coasts of South America. Due to the warm Brazil Current along the east coast, commercial fisheries for penaeid shrimps extend much farther south there than along the west coast, which is washed by the cold Peru Current. Solid arrows: cold currents; broken arrows: warm currents.

marketed and consumed locally. Potentially of more importance as an item of export from Chile is the *langostino,* which is shipped to the United States as "rock lobster tails." Yet the *langostino* of Chile is not a rock lobster. It is a member of the family Galatheidae, comprising lobster-like creatures closely allied to the hermit crabs and the king crab *Paralithodes camtschatica*. *Langostinos* are caught by otter trawls in deep water at about 500 feet, the most important species in the catch being *Cervimunida johni* and *Pleuroncodes monodon*. The latter, known as the red crab, has a very close relative living along the coast of Baja California (see Chapter 1).

Three

Distribution and Fisheries in Fresh Water and on Land

Marine species of shrimps, lobsters and crabs, including many described in the preceding chapter, are often more familiar to us than are species that live in fresh water and on land. We now examine some of these less familiar animals.

Freshwater Crayfishes

Among better known decapod crustaceans living in fresh water are the crayfishes. These animals look very much like true lobsters, so much so, in fact, that occasionally there have been reports of "lobsters" living in fresh water. In matter of fact, lobsters live only in sea water. But because of close structural similarities, freshwater crayfishes and true lobsters are classified together in the same infraorder Astacidea, but in separate superfamilies; the Astacoidea and Parastacoidea for freshwater crayfishes and the Nephropoidea for true lobsters.

Freshwater crayfishes of the genera *Astacus* and *Austropotomobius*, family Astacidae, are found from the British Isles eastward across Europe and into Asia, almost to the Ob River. All of these crayfishes occur between 60° and 40°N; that is, from the latitude of southern Norway,

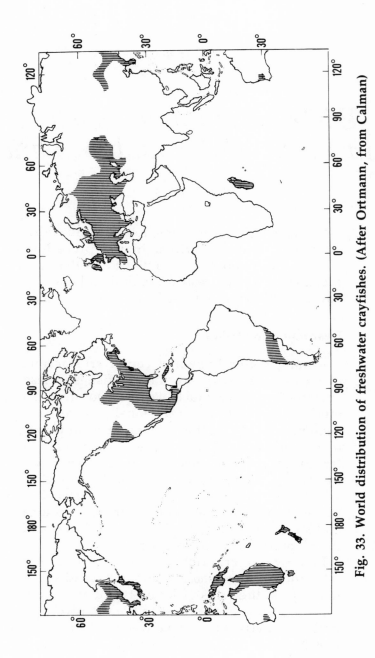

Fig. 33. World distribution of freshwater crayfishes. (After Ortmann, from Calman)

Sweden, and Finland to that of northern Italy, Spain, and Greece. Many species are considered delicacies for the dinner table. A single genus, *Cambaroides*, of the subfamily Cambaroidinae, family Cambaridae, is found in eastern Asia in the Amur Basin, in Korea, and in Japan.

Two large areas of the United States and Canada are inhabited by freshwater crayfishes. West of the Continental Divide from British Columbia southward to California and in Idaho and Utah, and east of the Divide in certain headwaters of the Missouri River within Montana and Wyoming, there are five species of *Pacifastacus*, all members of the family Astacidae. They are relished as food and are marketed in the Pacific Northwest. These crayfishes have been introduced into Sweden and Japan.

Central and eastern portions of the United States and southern Canada are inhabited by more than 230 species and numerous subspecies of freshwater crayfishes, almost all of which belong to the subfamily Cambarinae, family Cambaridae. Commercial fishing for these crayfishes takes place along the Mississippi River and on the Gulf Coast.

In Louisiana, "crawfishing," as it is often called, is a sizable industry. The species taken, usually in traps, are the red, or swamp, crayfish, *Procambarus clarkii*, and the white, or river, crayfish, *Procambarus acutus acutus*. Crawfishing is also outdoor recreation for many a picnicker, who generally catches the animals in baited nets set in the shallow waters of a swamp. Children often capture crawfish by tying a piece of meat to the end of a string and pulling the animals out, two or three at a time, as they cling to the bait.

A knowledgeable crawfish eater separates the "head" of the boiled animal from the tail, and, by simultaneously crushing the shell and sucking on the open end, draws out first the tasty "fat" from the head and then the succulent meat from the tail. Crawfish may also be eaten as a bisque, which, in canned form, can be obtained in various parts of the country.

In New York City, Scandinavian specialty shops and restaurants feature a variety of crayfish dishes, particularly during Kräfttidden, or "crayfish time," which runs from early August to the end of September, when crayfishes are said to be at their tastiest. In Sweden, on the first day of Kräfttidden, more than one and one-half million crayfishes are reported to be eaten in Stockholm alone.

From approximately 15°N to the equator, no freshwater crayfishes are to be found anywhere in the world. Then in New Guinea and Australia one encounters the genus *Cherax* of the family Parastacidae, superfamily Parastacoidea. Other genera of this family occur in Australia, Tasmania, New Zealand, Madagascar, and the southern part of South America.

Some species of Australian freshwater crayfishes, notably *Astacopsis gouldi, Cherax tenuimanus,* and several species of the genus *Euastacus,* get very large, with adult specimens reaching an average length of one foot or more. *Astacopsis gouldi,* known as the giant Tasmanian crayfish, may attain 16 inches in length and weigh up to eight pounds; it is the largest of all known freshwater crayfishes. *Euastacus armatus,* which reaches six pounds in weight and is known in Australia as the "Murray River lobster," is fished commercially on a small scale. It has been rated by gourmets as among the finest-flavored crustaceans.

Freshwater Shrimps and Crabs

Shrimps are abundant in fresh water, and fisheries for them exist in many countries. This is particularly true in Asia, where large quantities of freshwater shrimps are caught daily by local villagers in rice paddies, lakes and canals.

One genus of freshwater shrimp, *Macrobrachium,* is pantropical in distribution, occurring primarily in areas of the world where, except for New Guinea, freshwater crayfishes are lacking. Twenty-eight species of *Macrobrachium* are found in North and South America, six of them in the United States. Like others of this genus, all are recognizable by their second pair of thoracic legs, which are long and slender and terminate in a claw. The generic name *Macrobrachium* means "long-armed."

Several American species of *Macrobrachium* are very large. In the United States, among the largest of these "river shrimps" ever recorded was an individual of *Macrobrachium*

Fig. 34. Large freshwater shrimp known scientifically as *Macrobrachium carcinus*. **Claws of this specimen were banded for safety and ease in handling. Meat in tail offers delectable feast.**

carcinus that measured nine inches in length, not including the claws. A smaller species, *Macrobrachium ohione*, has for many years been trapped commercially in Louisiana.

A genus of freshwater shrimps found commonly in the United States is *Palaemonetes*, one species of which, *P. paludosus*, is economically very important not as a food for man—the shrimp is too small for this—but as a vital link in the food chain that supports many game fishes and other fishes marketed commercially. The shrimp is abundant in quiet water, where it clings to emergent vegetation and broken branches, as well as to the floating plant known as duckweed, which often covers the surface of a lake or pond. Should floods occur, the shrimp may be found among inundated grasses and bushes.

Freshwater crabs are of common occurrence in warm, humid parts of the world. A continuous band drawn around the globe between 35°N and 35°S would contain geographic areas in which freshwater crabs are to be found. Discontinuities would, of course, occur where there are oceans or arid regions such as the Sahara Desert and the plateaus of Saudi Arabia, Iran, and Afghanistan.

Although all freshwater crabs used to be considered as members of one family, the Potamonidae, they were subsequently reassigned to three families and, quite recently, to nine families. For the purposes of this book, we can use the three-family designation. The family Potamidae consists largely of crabs of the genus *Potamon*. These are the freshwater crabs of the Old World and, in all, several hundred species of this genus are distributed throughout large areas of Europe and Asia. The families Pseudothelphusidae and Trichodactylidae comprise the freshwater crabs of the New World. Pseudothelphusids are abundant in Mexico, the West Indies, Central America, and northern South America. Except for one Central American and one or two Mexican species, trichodactylids are found only in South America.

Almost any kind of fresh water seems acceptable to freshwater crabs. They live not only in slowly flowing rivers close to the sea, but also in rushing mountain streams and mountain waterfalls that are six to eight thousand feet above sea level. Freshwater crabs can be found in little brooks, in ditches, in irrigation canals of rice paddies, and in small lakes, ponds, miry pools, and swamps. The crabs are found where the substratum is of gravel, sand, mud, or moss.

In the past at least, even cities were not without their freshwater crabs. The American taxonomist Dr. Mary J. Rathbun, in her treatise entitled *Les Crabes d'Eau Douce*, published in Paris in 1904–1906, reported that freshwater crabs were to be found in the canals and reservoirs of Calcutta and in the basins of the fountains of Rome.

During the dry season, freshwater crabs may be taken from dry river beds and from gorges, pits, deep forests, and open fields. To protect themselves, the crabs in these areas dig burrows or hide beneath roots and rocks or stones, and sometimes in rotten wood. In their ability to live for long periods out of water, certain freshwater crabs resemble land crayfishes and land crabs.

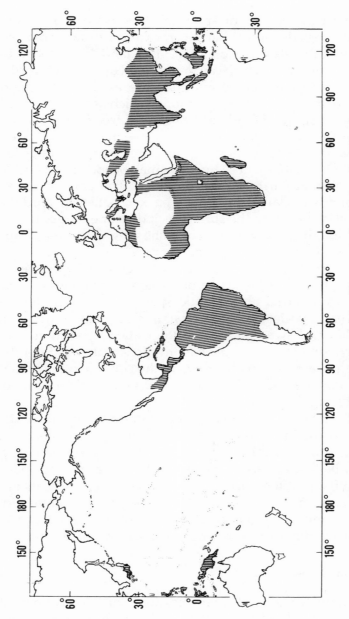

Fig. 35. World distribution of freshwater crabs. (After Rathbun)

Land Crayfishes

No true lobsters or spiny lobsters live on land, but some crayfishes do. A few hermit crabs and a number of species of true crabs also live on land. The land habitat has certain disadvantages, which can be summed up in a few words—it is dry and its temperatures often are extreme. Yet land also offers many advantages, among which are abundance of oxygen, variety of food and cover, and freedom of movement.

Truly terrestrial crayfishes live in wet pastures and meadows, where they dig deep burrows reaching down to the water table. These burrows may be located by the "chimneys" that surround the entrance. The chimneys are hard mounds of clay formed from packed pellets of mud that the crayfish brings to the surface as it digs or enlarges its burrow. Although usually about six inches high, chimneys up to 18 inches in height have been seen.

In Australia and Tasmania the genus *Engaeus* is almost entirely terrestrial, while other genera of crayfishes contain both terrestrial and aquatic species. In the United States, several species, notably *Cambarus diogenes,* burrow in wet fields and swampy areas. Other species, such as the mud crayfish, *Orconectes immunis,* live in streams and ponds most of the year, but dig burrows in the soil when the water level falls or the body of water dries up during late summer and autumn.

Aquatic species of crayfishes can do considerable damage by burrowing into the retaining walls of channels and dams and into the banks of rivers and streams. Terrestrial species can be just as serious pests. In Australia, for instance, land crayfishes often burrow in orchards under the roots of trees. In some meadows the crayfishes make large community burrows and cause the ground to be so weakened that it may collapse under the weight of cattle and horses.

Land Crabs

In the tropics and subtropics, and particularly on innumerable coral islands of the Pacific, there are species of true crabs that are commonly referred to as "land crabs." In some instances, these crabs are no more terrestrial than are many species of freshwater crabs that are able to remain out of water for long periods of time. A principal difference between the two groups lies in the type of water to which they return when ready to release their spawn, with freshwater crabs seeking fresh water and land crabs seeking the ocean at this time.

Most conspicuous of land crabs are those of the genus *Cardisoma* in the family Gecarcinidae. These are big animals, having long, spiny legs and a body that may be four or five inches wide. In an adult male one claw is very large.

Of the genus *Cardisoma*, the species best known to Americans is the great land crab, *C. guanhumi*, which is found from Bermuda, southern Florida, the Bahamas, and Texas to Brazil. The crab feeds on the meat of fallen coconuts, mangrove leaves, and crops such as rice, tomatoes, and sugar cane. Yet it is also an excellent scavenger, helping to clean up edible debris and thus improving sanitation in congested areas.

There are fisheries for *Cardisoma guanhumi* in various parts of its range. In Puerto Rico, for example, a fishery for the juey,

Fig. 36. Great land crab of southern Florida, Bahamas, and West Indies. Vegetarian, yet also scavenger, this crab serves as choice item of food, with fatty "liver" and meat from large claws the favored portions.

or cangrejo, as the crab is called there, has existed for many years, and numerous Puerto Ricans use it as a principal source of income. Although often employing wooden traps, the fishermen have several other methods for capturing the crab at or in its burrow. During the rainy season, when the crabs are easier to catch because they leave their burrows during the day, amateur crabbers join the chase. The fatty "liver" and meat from the claws are choice portions of this animal.

On the west coast of the Americas, from Baja California to Peru, there lives another species of land crab, *Cardisoma crassum*, which differs very little in appearance from *C. guanhumi*. These crabs are known as twin, or geminate, species, since they are more closely related than are ordinary species. It is believed that at one time the Atlantic and Pacific Oceans were confluent and that one species of *Cardisoma* occupied a continuous range. When the oceans became separated from each other, the Atlantic and Pacific forms developed differences in structure, which, though slight, are significant enough for scientists to consider the Atlantic and Pacific forms separate species. Twin species are found also among other crabs and in other groups of animals, notably echinoderms and fishes.

Fig. 37. Black land crab, inhabitant of southern Florida, Bahamas and West Indies, and a desired food item locally where abundant.

A second branch of the family Gecarcinidae is that of the genus *Gecarcinus*. Five species occur in the Americas. Along the east coast, the black, or blue, land crab, *Gecarcinus ruricola*, is common in southern Florida, the Bahamas, and the West Indies. This crab approaches *Cardisoma guanhumi* in size but is more striking in coloration. In regions where *G. ruricola* is abundant, it is captured and marketed locally. It can be plentiful in the markets of Nassau.

Smaller than *Gecarcinus ruricola*, but almost as colorful, is the purple land crab, *Gecarcinus lateralis*. This crab occurs from Bermuda, Florida, Texas, and the Bahamas to Colombia and Venezuela. It is purple, or purple and plum colored, hence its common name. In some areas, the crab, which is too small to be an important item of food, can be a serious agricultural pest. The crab burrows not only on the upper beaches and in the dry grasslands nearby, but also in lawns, gardens, pastures, and cultivated fields, where it feeds on roots and leaves of crops. Yet the crab is also an excellent laboratory animal, useful in the study of many physiological processes, a fact that, in the eyes of many scientists, offsets its drawbacks as a pest.

Often referred to as land crabs are terrestrial hermit crabs of the genus *Coenobita*, family Coenobitidae. They are found in huge numbers on the land areas of all atolls. Like their marine relatives, terrestrial hermit crabs keep their soft abdomen tucked away within a borrowed mollusk shell, which they drag around with them wherever they go. During the day they seek protection from sun and heat by hiding under branches, leaves, rocks, and other cover, and at night they emerge to scavenge for food.

Like terrestrial species of true crabs, terrestrial hermit crabs are restricted to tropical and subtropical coasts and islands, where temperatures are favorably warm, moisture is plentiful, and the ocean, in which to release their larvae, is close at hand. In the Bahamas and southern Florida, the species *Coenobita clypeatus* is abundant. The range of this species extends southward to Venezuela.

Terrestrial hermit crabs perform a useful function in eating dead organic matter that otherwise would decay and become

Fig. 38. Land hermit crab of the species *Coenobita clypeatus,* which is abundant in southern Florida and Bahamas, is adept at climbing trees to obtain young, tender leaves to eat, and has been found to make an interesting pet that accepts a variety of plant or animal food. (Courtesy of Jacques van Montfrans)

a health hazard. But their scavenging habits sometimes make them a nuisance. With their strong legs they can climb easily, and with their massive claws they can break open flimsy storage containers of food, as some field biologists have found to their sorrow. In recent years, terrestrial hermit crabs, at least of the species *clypeatus,* have become popular as pets. They can be kept in a covered terrarium with branches on which to climb and moist sand in which to burrow at the time of molting. A shallow dish of fresh or sea water is necessary for drinking. The behavior of land hermit crabs is very interesting and easy to observe under such conditions.

A close relative of the terrestrial hermit crab is the coconut crab, *Birgus latro*. This large animal is present on most islands of the tropical Pacific and western Pacific and on islands of the Indian Ocean that lie south of the equator. Its flesh is tasty and eagerly sought. Its diet, which is varied, includes the meat of coconuts. The crab can climb coconut palms, but

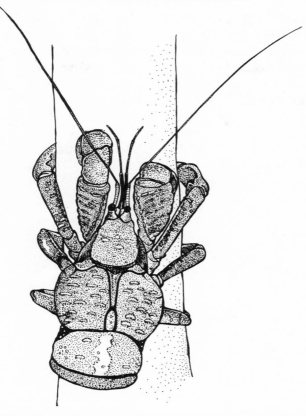

Fig. 39. Coconut crab, inhabitant of tropical islands and close relative of the land hermit crab, its flesh tasty and eagerly sought.

whether, as some persons claim, it can husk a coconut and cut its stalk, later to climb down the tree and eat the white meat of the fallen and broken nut, is open to question. There is no evidence that a coconut crab can open an undamaged nut.

Coconut crabs are shy creatures and leave their burrows only at night. According to the crustacean specialist Dr. L. B. Holthuis, the natives of New Guinea catch these crabs by placing coconut and other bait near the burrows in the evening and waiting till the crabs appear. Then the natives press the crabs against the ground and tie their claws, thus rendering them harmless.

An even more original method of catching coconut crabs is described by Dr. Holthuis. After a crab has climbed a coconut palm, a native of New Guinea attaches a wreath of grass to the trunk of the tree as far up as possible. When the crab comes down the tree and touches the grass with its abdomen, it lets go of the tree trunk and falls to the ground. Using a similar method, but employing leaves instead of grass for the "wreath," the natives of the Celebes place stones at the base of the tree so that the falling crab will be smashed when it hits the ground.

Four

Structure and Function

To a scientist, structure is a primary basis upon which to classify animals. This is apparent in Chapter 1. To an animal, structure is the key to function. In this chapter we examine shrimps, lobsters, and crabs for ways in which the structure of various tissues and organs and the workings of various systems enable the animal to carry on its vital functions.

Shrimps

As a decapod crustacean, the white shrimp, *Penaeus (Litopenaeus) setiferus,* is rather primitive. It has been selected for use here because it illustrates not only the structure of a shrimp, but also the generalized body plan of a decapod crustacean.

A large portion of the white shrimp, as of any other shrimp, consists of muscle and shell, or exoskeleton. In fact, the largest of the three natural divisions of the body; namely, the abdomen or "tail," consists of little other than muscle and shell. Two main masses of muscle are the "meat" of the shrimp's tail: (1) the relatively small dorsal abdominal muscles, which lie above the intestinal tract, or gut, and above

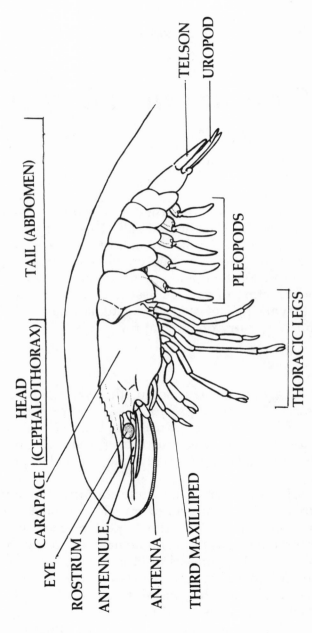

HEAD

CARAPACE (CEPHALOTHORAX)

TAIL (ABDOMEN)

EYE

ROSTRUM

ANTENNULE

ANTENNA

THIRD MAXILLIPED

THORACIC LEGS

PLEOPODS

TELSON

UROPOD

Fig. 40. White shrimp, *Penaeus setiferus*, in lateral view. Appendages of left side only are shown. (After Rathbun, also Young)

the dorsal abdominal artery, both of which are removed in preparation for eating, and (2) the large ventral abdominal muscles, which extend from either side of the intestinal tract and dorsal abdominal artery ventrally to both sides of the ventral abdominal nerve cord.

For swimming quietly, the white shrimp uses its five pairs of abdominal appendages, known as pleopods. But when the shrimp moves rapidly, it does so by contracting its ventral abdominal muscles and curving forward its tail fan, which is composed of a centrally situated telson and a pair of lateral appendages known as uropods. The powerful thrust exerted by tail and tail fan upon the water propels the shrimp backward with extraordinary speed. The tail of the shrimp returns to its normal, more or less elongated, position by the contraction of the dorsal abdominal muscles, which act as extensors. The tail's flexibility results from deep folds of thin, soft chitin that link the six segments of the tail to one another.

Within the cephalothorax of the white shrimp are large portions of the digestive, circulatory, nervous, and reproductive systems. The long digestive tract, or gut, has three main subdivisions known, respectively, as foregut, midgut and hindgut. Food particles picked up by the mouth parts are ground by the mandibles and swallowed, whereupon they enter the narrow, tubular, muscular esophagus, which is the initial portion of the foregut. Lined with chitin, the esophagus nonetheless can accommodate large amounts of food since it has one anterior and two lateral folds loosely filled with connective tissue. When these folds become unfolded, the esophagus can distend greatly.

From the esophagus, food particles enter the anterior chamber, the second portion of the chitin-lined foregut. Many authors have called this the cardiac stomach. The anterior chamber has lateral longitudinal folds that permit it to expand when filling with food. The anterior chamber also has ventrally situated longitudinal ridges that lead back to the openings of the midgut glands in the caudad part of the posterior chamber, frequently called the pyloric stomach.

Walls of the anterior chamber contain a triangular structure consisting of a median tooth and a row of tooth-like denticles

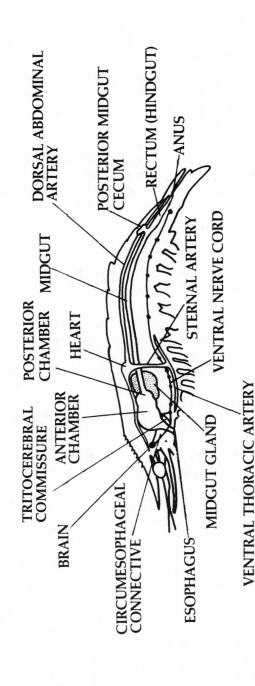

TRITOCEREBRAL COMMISSURE

POSTERIOR CHAMBER

MIDGUT

DORSAL ABDOMINAL ARTERY

POSTERIOR MIDGUT CECUM

RECTUM (HINDGUT)

ANUS

BRAIN

ANTERIOR CHAMBER

HEART

STERNAL ARTERY

VENTRAL NERVE CORD

CIRCUMESOPHAGEAL CONNECTIVE

ESOPHAGUS

MIDGUT GLAND

VENTRAL THORACIC ARTERY

Fig. 41. Internal anatomy of white shrimp, showing in lateral view major organs of digestive, circulatory and nervous systems.

along each side. When food enters the anterior chamber, the muscles that insert on the chamber alternately contract and relax, thereby causing the median tooth to move against the denticles and lateral ridges. In so doing, this grinding apparatus, termed the gastric mill, breaks down the food into very fine particles.

While food is within the anterior chamber, it is mixed with digestive juices that flow forward ventrally from the posterior chamber. The juices enter the caudad part of the posterior chamber via ducts that originate in the lumen, or cavity, of many-branched tubules constituting the paired midgut glands. Thus, the lumen of the tubules is continuous with the lumen of the gut.

Digestion of food takes place partly in the anterior chamber, partly in the posterior chamber, and partly in the tubules of the midgut glands. In the posterior chamber, there is a filter formed by two lateral ridges and one ventral median ridge densely covered with hair-like setae. Owing to this filter, only fluid and minutely divided food particles can pass from the posterior chamber into ducts leading to the midgut glands and thence into its branching tubules for further digestion. From the midgut glands, end products of digestion are readily absorbed into the hemolymph.

Fine indigestible material within the midgut glands is forced back into the posterior chamber and then into the straight, unlined, tubular portion of the midgut. Here end-products of digestion enter the hemolymph via the many small blood vessels connecting the tubular portion of the midgut with the dorsal abdominal artery just above. Here also the fine indigestible material is mixed with larger indigestible particles that had been filtered away from the openings of the midgut glands and had passed directly from the posterior chamber into the tubular part of the midgut.

Within the midgut, indigestible material is packaged into long fecal pellets and enclosed within a membrane, the peritrophic membrane (from the Greek, *peri*, around; *tropho*, feed), which is secreted by epithelial cells of the midgut and is mucoid in nature. Strong peristaltic contractions of the midgut push the fecal pellets along to the chitin-lined hindgut, which is enlarged as a rectum. A series of rapid contractions by the rectum then forces the fecal pellets out of the body by way of the anus.

At the junction of midgut and hindgut in the sixth abdominal segment, the midgut gives rise to a diverticulum, called the posterior midgut cecum by some authors and the hindgut or rectal gland by others. The function of this organ is not known, but its cells appear to be secretory.

Presumably the midgut would be distinguishable from the foregut and hindgut by its lack of chitinous lining. Thus, the esophagus, anterior chamber, and cephalad portion of the posterior chamber are lined with chitin and are clearly foregut. The caudad portion of the posterior chamber also is lined, although incompletely, with chitin; yet this is midgut. Dorsally and laterally, the chitinous lining of the foregut in the white shrimp (and in many other decapod crustaceans) extends into the midgut well past the openings of the midgut glands; ventrally a caudad extension of the chitinous lining separates the openings of the midgut glands and also covers the epithelium of more posterior portions of the midgut for some distance. These caudad extensions of the chitinous lining probably direct sand and other indigestible particles to the peritrophic membrane for packaging without damage, en route, to the delicate area around the openings of the midgut glands.

The heart of the white shrimp has three pairs of small openings known as ostia. Through these ostia, the blood flows into the heart from the surrounding area, which is termed the pericardial sinus, or pericardium. Valves prevent the blood from leaking out through the ostia as the heart contracts. Instead, the blood is driven into major arteries, most of which run forward to supply blood to the sense organs and to vital organs within the cephalothorax. However, the sternal artery runs toward the ventral region of the shrimp, where it gives rise to a ventral thoracic artery that supplies blood to the thoracic appendages and to the thoracic portion of the ventral nerve cord. The dorsal abdominal artery leaves the heart posteriorly and supplies blood to the gut, the abdominal muscles, and the abdominal portion of the ventral nerve cord.

The nervous system of the white shrimp consists of a brain (supraesophageal ganglion), which is situated dorsally in the head, two circumesophageal connectives that pass on either side of the esophagus and are connected with each other by the tritocerebral commissure, and a ventral nerve cord, which runs posteriorly the entire length of the shrimp and at more or less regular intervals is swollen into bulbous ganglia. The entire central nervous system of the white shrimp, as of other decapod crustaceans, is fundamentally "ladder-type" in structure, but in most regions the two longitudinal halves of the "ladder" have fused. As a consequence, the word "ganglion" generally refers to a pair of laterally fused ganglia.

The brain receives nerves from sense organs of the head, notably the eyes and antennae, and supplies nerves to the muscles that operate these sense organs. In the ventral nerve cord, the first ganglion (subesophageal ganglion) and the remaining ventral ganglia (five in the thorax and six in the abdomen) receive nerve fibers from sensory cells widely dispersed through the body of the shrimp and supply nerves to muscles that move the mouth parts, thoracic legs, pleopods, and tail.

In addition, lying on the circumesophageal connectives is a pair of connective ganglia, or stomatogastric ganglia (stō-mă-tō-GĂS-trĭk; from the Greek, *stoma,* mouth; *gaster, stomach*). The connective ganglia and the stomadeal ganglion on the anterior surface of the esophagus combine to form the stomadeal system, which supplies nerves to the esophagus and the foregut.

In the forward part of the cephalothorax of the white shrimp, situated on the second, or antennal, segment are the kidneys, which because of their location are often called antennal glands. Each kidney is made up of a small dorsal portion that lies above the brain and a large ventral portion lying beneath the brain. The two portions of each kidney are connected with each other by lateral arms. Part of the ventral portion extends into the antenna on the same side of the animal. A short duct from this portion of each kidney leads to the exterior through an excretory pore, which lies at the base of the antenna on its inner (medial) side. In higher shrimps, the Caridea, a bladder also is present.

There are 19 pairs of gills in *Penaeus setiferus*. Three pairs occur in each thoracic segment, except the first and last, where there is one pair. In any given segment the gills may be attached to the base of the limb, to the flexible membrane between limb and body, or to the body wall. A gill of the white shrimp consists of a primary supporting axis known as a rachis (RAY-kĭs; from the Greek, *rhakhis,* spine, or backbone), from which secondary supporting structures emerge at right angles. On the secondary supporting structures are many gill filaments that in turn protrude at right angles. Each secondary supporting structure with its attached filaments nests against the preceding one. In caridean shrimps, the gill filaments are flattened and plate-like and protrude directly from the primary supporting structure; this type of gill also is found in crabs.

In shrimps, as in all other decapod crustaceans, the gills lie within two branchial chambers, each of which results from a deep lateral fold of the carapace. The beating of a leaf-like flap, the gill bailer, or scaphognathite (skă-FŎG-nă-thīte), causes water to enter the branchial chamber from below and behind; that is, through openings between the thoracic legs and in front of the abdomen. The water leaves the branchial chamber through a channel, directed toward the head, in which lies the beating gill bailer. As the water circulates through the branchial chamber, an exchange of gases takes place between the water and the blood in the gill filaments. At the same time there is a discharge of excess salts from the blood into the water and an uptake of needed salts from the water into the blood.

In the white shrimp the most conspicuous components of the female reproductive system are two ovaries that extend, partially fused, from the anterior region of the foregut posteriorly to the tail fan. The portion of each ovary that is within the cephalothorax consists of a forward-projecting lobe, which lies close to the esophagus and chambers of the foregut, and seven finger-like lateral lobes, which are situated above the midgut gland and beneath the heart. This arrangement makes the heart resemble a saddle straddling the

ROSTRUM EYE FOREGUT HEART

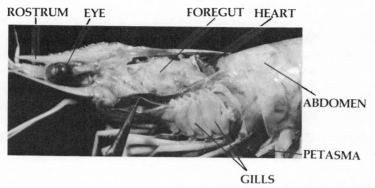

ABDOMEN

PETASMA

GILLS

Fig. 42. Preserved, partially dissected specimen of white shrimp, showing gill bailer (at tip of forceps) and gills in left branchial chamber, also petasma (male copulatory organ).

HEART ABDOMINAL LOBE

OVARY

FORWARD-PROJECTING
LOBE

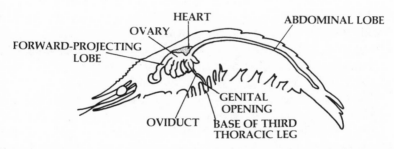

GENITAL
OPENING

OVIDUCT BASE OF THIRD
THORACIC LEG

Fig. 43. Female reproductive system of white shrimp, lateral view. (After King)

ovaries. The abdominal portion of the ovaries consists of two lobes, lying above and to the sides of the intestine and below and to the sides of the dorsal abdominal artery.

Emerging from each ovary at the sixth lateral lobe is an oviduct. Coursing ventrally, each oviduct opens to the exterior at a genital pore situated medially on the basal segment of the third thoracic leg. The opening is concealed within an ear-shaped protuberance covered with setae.

Externally and posterior to the genital openings of the female lies a structure that is adapted for receiving a packet of sperms, or spermatophore (spêr-MĂ-tō-phôre; from the Greek, *sperma*, seed; *phoros*, bearing) from the male during mating. Known as the thelycum (THĔL-ĭ-cŭm; from the Greek, *thelys*, female), this structure consists of several lobes and protuberances bearing stiff bristles.

The male reproductive system of the white shrimp includes a pair of partially fused testes that lie in a position quite similar to that of the ovaries in the female. Each testis has an anterior lobe projecting forward over the chambers of the foregut and six lateral lobes that lie over the midgut gland and under the heart. In place of a long abdominal lobe as in an ovary, each testis has a short posterior lobe.

A pair of ducts known as the vasa deferentia emerge from the main axis of the testes at their posterior margin, course ventrally, and open to the exterior at the genital pores situated medially on the basal segment of the fifth pair of thoracic legs. Each vas deferens has four distinct regions: a short, narrow proximal portion; a thickened, doubly flexed medial portion; a long, narrow tubular portion; and a much dilated, muscular terminal ampoule. Within the terminal ampoule the spermatophore is formed.

The spermatophore of the white shrimp roughly resembles a pod. It consists of two halves, each of which contains sperms enclosed within a sheath and surrounded by chitin. The thoracic legs of the male shrimp presumably assemble the spermatophore immediately after each half is expelled from the terminal ampoule of the corresponding vas deferens. The

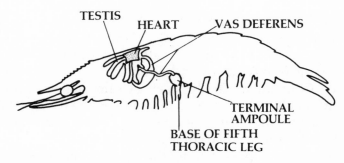

Fig. 44. Male reproductive system of white shrimp, lateral view. (After King)

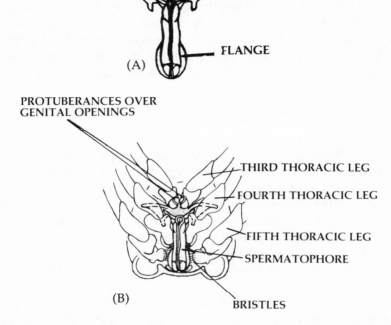

"WINGS"

FLANGE

(A)

PROTUBERANCES OVER
GENITAL OPENINGS

THIRD THORACIC LEG

FOURTH THORACIC LEG

FIFTH THORACIC LEG

SPERMATOPHORE

(B)

BRISTLES

Fig. 45. Spermatophore of male white shrimp as seen unattached (A) and attached (B) to female's thelycum on her ventral surface. "Wings" on spermatophore are securely anchored in groove between third and fourth thoracic legs, while bristles on base of fourth and fifth thoracic legs overlap flanges of spermatophore. (After King)

legs place the spermatophore within the trough of the petasma (pě-TĂZ-mǎ, from the Greek, *petasma,* spread out), a structure that results from modification of the first pair of pleopods. The petasma consists of stiffened longitudinal rods and folds of soft chitin that, when unfolded, result in a broadly inflated male copulatory organ.

During mating, the male shrimp uses the petasma to thrust the spermatophore against the thelycum of the female. Here bristles on protuberances of the thelycum overlap the spermatophore, thus helping to secure it. Two lobes known as "wings" on the spermatophore become anchored in a groove on the ventral surface of the female between her third and fourth thoracic legs. Despite these devices for securing the spermatophore, it is easily dislodged, and spermatophore-bearing females of white shrimp are not commonly caught in shrimp trawls. A pair of light-colored, pad-like structures situated just posterior to the thelycum are believed to play no role during impregnation.

Lobsters

In general body plan a lobster does not differ greatly from a shrimp. A lobster has the same type of muscular abdomen, or tail, which undergoes sudden flexion by contraction of the large ventral abdominal muscles and more leisurely extension by contraction of the smaller dorsal abdominal muscles. As in shrimp, the tail of a lobster provides the animal with its surest means of escape—jetlike propulsion backward.

In lobster as in shrimp, a carapace covers the head and the thorax and, except for the presence of the cervical groove, obscures the boundary between these two regions. In lobsters, the cephalothorax is commonly called the "body," while in shrimp this same region is known as the "head."

Despite similarities in body plan, it is quite easy to distinguish a lobster from a shrimp. Even as an adult, a shrimp is relatively small and its shell is somewhat fragile. An adult lobster, on the other hand, may reach very large size and acquire an extremely hard shell. Furthermore, a lobster is compressed dorsoventrally (from top to bottom), not laterally (from side to side), as is a shrimp. True lobsters have yet another distinguishing characteristic: their first pair of thoracic legs is modified as large claws, or chelipeds. In some species, such as the American lobster, *Homarus americanus*, one large claw, the crusher, is much heavier than the other claw, known as the pincer, or the biting, cutter, or ripper claw. The crusher of the American lobster occurs about as frequently on the right side of the body as on the left. These large claws are lacking in the spiny, or rock, lobsters.

As in a shrimp, many vital organs of a lobster are situated under the carapace within the cephalothorax. Here are the chitin-lined foregut, at least a portion of the midgut, and the midgut glands. Here also lie the brain, heart, gills, excretory organs, and a large part of the male and female reproductive organs.

In both true lobsters and spiny lobsters, the gastric mill is more highly developed than in the white shrimp. The gastric mill of the lobster is largely restricted to the region of the foregut in which the large, thin-walled anterior chamber gives way to the much smaller, thick-walled posterior chamber. At the constriction between the two chambers, three movable teeth, one median and two lateral, are attached to small, hard skeletal plates known as ossicles. These teeth chew the food, which arrives in the anterior chamber as long,

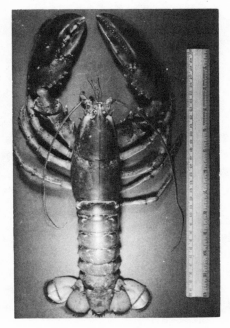

(A)

Fig. 46. (A) Male American lobster, showing characteristically heavy body, very large claws, and long, swollen carapace. (B) Female American lobster, showing somewhat lighter body, less massive claws, and shorter, unswollen carapace.

(B)

stretched, but unchewed pieces. A well-developed gastric mill is a useful device enabling a decapod crustacean, when safely hidden from its enemies, to chew its food at leisure, after having swallowed it in large pieces. The gastric mill is least developed in such decapod crustaceans as the shrimps, in which the mouthparts chew the food quite thoroughly before the food enters the esophagus.

In the walls of the anterior chamber, a lobster has many ossicles in addition to those of the gastric mill. These additional ossicles serve as a place of attachment for muscles that move the foregut and thereby enable the ossicles of the gastric mill to grind the food. Once the food has been ground thoroughly, it passes through a setose filter that prevents all but the finest particles from entering the midgut glands through ducts that open into the unlined caudad portion of the posterior chamber.

The midgut of the American lobster is long, extending back

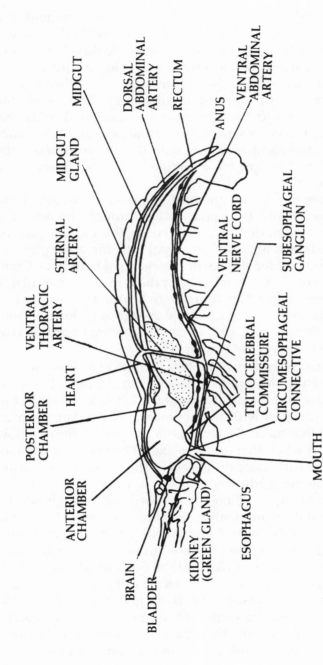

Fig. 47. Internal anatomy of the American lobster, showing major organs of the digestive, circulatory, nervous, and excretory systems.

to the last abdominal segment, where it connects with the chitin-lined hindgut, which has become modified as an enlarged rectum. A posterior midgut diverticulum, or cecum, arises just in front of the junction of midgut and rectum. Undigested wastes are egested from the rectum through the anus. In spiny lobsters, the midgut is very short, while the hindgut is long and contains many longitudinal folds. No enlarged rectum is present, the terminal portion of the hindgut being narrow and very muscular. By their contraction the muscles of the hindgut force undigested (fecal) material out through the anus.

In the foregut, midgut, and midgut glands of the American lobster, digestion of food takes place through the action of digestive enzymes that are secreted by the midgut glands. These glands are also the principal site for absorption of digested food and for storage of reserve food materials. Chefs call the midgut glands of the lobster the tomally; accumulated food reserves make the tomally rich and flavorful when cooked. The tomally can easily be recognized, for it is soft, large, and many-lobed, and, in color, is green, bright yellow, yellow-green, or yellow-brown.

The circulatory system of a lobster is not very different from that of the white shrimp. In lobsters as in shrimps, the heart lies under the middorsal surface of the cephalothorax just in front of its junction with the abdomen. Three pairs of ostia allow blood that has collected within the pericardium to flow into the heart when that organ relaxes. During contraction of the heart, the ostia close and prevent the blood from flowing back into the pericardium.

From the heart, the blood flows forward through several arteries to vital organs within the cephalothorax. The blood also flows into the dorsal abdominal artery and its paired branches in each segment. These supply blood to the ventrally situated flexor muscles and the dorsally situated extensor muscles of the abdomen. A sternal artery carries blood to the gonads, then courses ventrally to give rise to the ventral thoracic artery and the ventral abdominal artery. In lobsters, as in shrimps, the ventral thoracic artery carries blood to most thoracic appendages and to the thoracic portion of the ventral

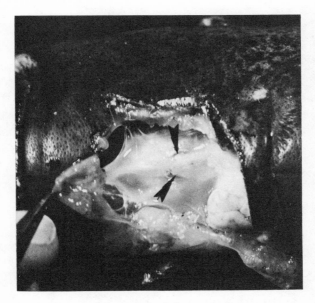

Fig. 48. Beating heart of American lobster, as it appears suspended within pericardium. Blood enters heart from pericardium via ostia (arrows). Abdomen of lobster is at left.

nerve cord. In lobsters, but not in shrimps, the ventral abdominal artery extends through the abdomen, supplying blood to the last two pairs of thoracic legs, the ventral nerve cord, the posterior part of the hind gut, and the tail fan.

The central nervous system of the American lobster differs little from that of the white shrimp. Lobsters, like shrimps, have a brain, or supraesophageal ganglion, composed of several fused paired ganglia. Running ventrally and posteriorly from the brain are two circumesophageal connectives, a slight swelling on each connective as it passes the esophagus marking the position of the stomatogastric, or connective, ganglia. Behind the esophagus, the connectives are joined by the small tritocerebral commissure.

Due to fusion of the first two thoracic ganglia with three cephalic ganglia to form the subesophageal ganglion, the thoracic portion of the ventral nerve cord in the American lobster contains only five additional ganglia (ganglia of the

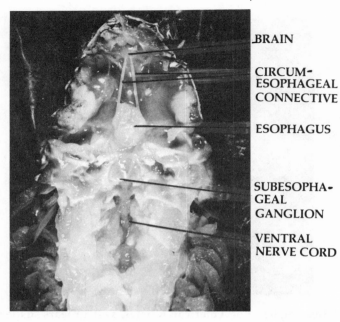

BRAIN

CIRCUM-
ESOPHAGEAL
CONNECTIVE

ESOPHAGUS

SUBESOPHA-
GEAL
GANGLION

VENTRAL
NERVE CORD

Fig. 49. Dorsal view of dissected American lobster, showing various parts of central nervous system.

last two thoracic segments have fused with each other). The abdominal portion of the ventral nerve cord contains six ganglia, one in each segment. This arrangement of thoracic and abdominal ganglia is similar to that in the white shrimp. In spiny lobsters, the thoracic ganglia have undergone greater fusion. Dr. C. J. George and his co-workers at Wilson College, Bombay, India, reported that in the thorax of *Panulirus polyphagus* there are only two ganglionic masses. The larger, anterior ganglionic mass has resulted apparently from fusion of nine pairs of ganglia (three cephalic, six thoracic), while the smaller, posterior ganglionic mass has come from fusion of two thoracic pairs. Yet in its abdomen, *Panulirus polyphagus* retains the original number of six ganglia.

As in shrimps, the brain of lobsters receives nerves from sense organs of the head, notably the eyes and antennae. Ganglia of the circumesophageal connectives supply minute nerves to the foregut. Ganglia of the thorax supply nerves to the mouth parts and thoracic legs. Abdominal ganglia furnish the nerve supply to flexor and extensor muscles of the abdomen, to the intestine, and to the abdominal appendages.

The kidneys of lobsters, like those of shrimps, are known as antennal glands. More compact than in shrimps, the kidneys of lobsters have a pale olive–green hue and thus are often called green glands. They lie on each side of the body, below and in front of the foregut. Urine that is formed in a glandular portion passes into tubes that enter a duct leading from a dorsally situated bladder. There is no direct connection between bladder and glandular portion, so the bladder can be filled only when urine backs up through this duct. Subsequently, the urine is released to the exterior via the same duct, which opens on the basal segment of the antenna.

Gills of lobsters are of a type known as trichobranch (TRĬK-ō-brănk; from the Greek, *thrix,* hair; *branchia,* gills), for they are composed of numerous filaments arranged, plumelike, around a central axis. As in shrimps, on any given thoracic segment there may be as many as four pairs of gills, one pair on the basal segment of the limbs, two pairs arising

Fig. 50. Left kidney (arrow), or green gland, of American lobster. Foregut appears above and to right, left eye above and to left.

from the soft membrane linking the limbs to the body, and one pair on the side of the body just above the limbs. In the American lobster the full complement of gills occurs at the base of the second, third, and fourth thoracic legs, with fewer pairs on the remaining thoracic segments except the first, which lacks gills. In all, there are 20 pairs of gills in the American lobster.

On each side the gills lie within the branchial chamber, which is formed, as in shrimps, by a deep lateral fold of the carapace. Access to the branchial chamber is through very small openings between the appendages and two larger openings, both ventral, one at the posterior end of the branchial

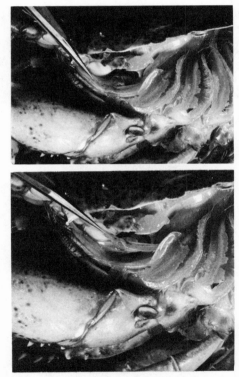

Fig. 51. Top, gill bailer of American lobster in lowered position (tip of forceps). Bottom, gill bailer partially raised. Gills in branchial chamber are visible at right.

chamber and the other at its anterior end. In a channel at the anterior opening is the leaf-like flap known as the gill bailer, or scaphognathite, which by its rapid beating drives water forward in the channel and out of the branchial chamber. At the same time the current thus established within the branchial chamber causes water to enter the ventral and posterior openings, principally the latter. Every few minutes, the gill bailer reverses its beat for a few strokes, thereby causing the current of water to flow in the opposite direction. By this reversal of current, silt and other debris that may have settled on the gills are loosened and can be flushed from the chamber.

Fig. 52. Dissected female American lobster, from above, showing heart (arrow) lying over ovaries, which run forward, under and to sides of foregut. When heart is removed (right), crossbar connecting ovaries is visible (tip of forceps). Ovaries extend well into abdomen.

In the American lobster the ovaries of the female appear in the form of a lettter H, with the cross bar at the forward margin of the heart and with longitudinal lobes extending forward and backward through much of the animal. The stage of ovarian development is apparent from the color, bright yellow or flesh-colored early in development, then salmon, light green, and finally a rich dark green by maturity. After cooking, the mature, egg-filled ovaries are bright red and are known as the coral.

From the ovaries, paired ribbon-like oviducts emerge at a level just below the heart, then quickly narrow as they run outward to the body wall and downward to the base of the third pair of thoracic legs, where they terminate on the inner surface of the basal segment. Externally and medial to these openings is a triangular, bluish structure extending from the base of the third to just beyond the base of the fourth pair of thoracic legs. This is the seminal receptacle, a small pocket in

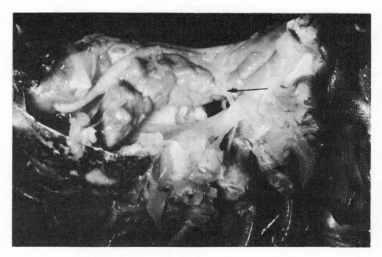

Fig. 53. Dissected female American lobster, lateral view. Long, slender ovary runs forward (to left), under and along side of foregut and back into abdomen. Oviduct (arrow) emerges from ovary just below heart (removed here) and courses downward to genital opening at base of third thoracic leg.

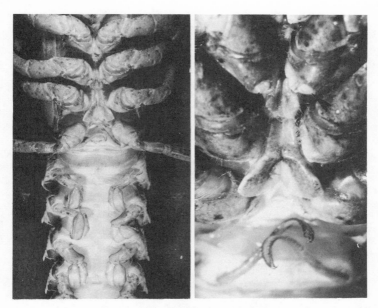

Fig. 54. Ventral surface of female American lobster, showing triangular seminal receptacle (enlarged at right) and small, soft first pair of abdominal appendages, or pleopods. Genital openings (arrows) lie at base of third pair of thoracic legs.

the exoskeleton that receives sperms from the male during mating.

In males of the American lobster the testes, which are pale tan–grey in color, may be H-shaped, like the ovaries of the female, or longitudinally paired, without a cross bar. From the testes, paired ducts, the vasa deferentia, emerge beneath the heart, at approximately the same place that the oviducts emerge from the ovaries. Like the oviducts, the vasa deferentia run outward to the body wall before turning downward. At this point they become S-shaped, with their posterior margin thickened and glandular, capable of secreting a gelatinous material that coats the sperms as they pass through the duct. The vasa deferentia then become briefly bulbous and muscular and, following this, narrow and thin-walled, forming an ejaculatory duct that opens at a papilla on the

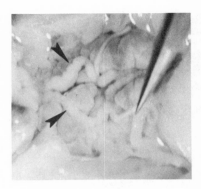

Fig. 55. Dorsal view (above) of dissected male American lobster, with testes (arrows) lying over midgut glands and under heart (removed here), and with paired vasa deferentia (tip of forceps) running outward to body wall and then downward, and terminating (below) at base of last (fifth) walking legs.

inner surface of the base of the fifth (and last) pair of thoracic legs.

The most obvious external difference between male and female American lobsters lies in the shape of the first pair of abdominal appendages. In the male these are the copulatory pleopods, relatively long, hard, grooved, and tapering. In the female these pleopods are small and soft. Yet there are other sexual differences, for mature males are heavier and have

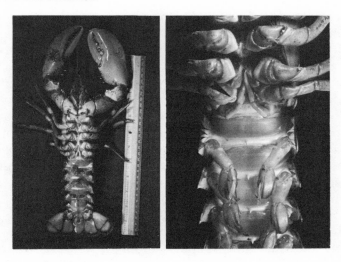

Fig. 56. Ventral view of male American lobster, showing (right) copulatory pleopods (first pair of abdominal appendages) which are long, hard, grooved and tapering. Adjacent to them lie genital openings, one of which (arrow) is clearly visible.

larger claws and a longer, more swollen carapace than have mature females.

Also in spiny lobsters, the sexes can be separated by differences in the abdominal appendages. In males the pleopods have one leaf-like terminal segment. In females the pleopods have two terminal branches, those of the first pleopods being leaf-like, while those of more posterior pleopods have one leaf-like branch and one rod-like branch used for attachment of eggs. In addition, the fifth pair of thoracic legs in male spiny lobsters terminates in a single, simple segment like that of more anterior legs, whereas in females the fifth pair of thoracic legs terminates in a small claw used in cleaning the attached eggs. Mature males tend to be larger than mature females.

Before leaving the subject of structure in American lobster, we may give some thought to coloration, for it can be surprisingly variable and frequently serves a protective function, enabling a lobster to blend with its background.

Normal, or "wild-type," coloration of American lobsters is mottled olive–green or dark blue–green above, with small black or green–black spots and often red tubercles and spines. On some lobsters the sides of the body and tail, as well as large portions of the claws, may be dusky orange, often dotted with green–black. Other lobsters are almost entirely dusky orange, with green–black spots. Such variations in color exist among lobsters of widely differing sizes, from the one-pound individuals commonly purchased in fish markets to the lobsters of 10 to 15 pounds or more that are caught on the southeastern part of George's Bank and in areas to the south.

American lobsters may be of other colors as well. Some, known as calico, or leopard, lobsters, are light yellow with purple–blue marbling or spots. Other lobsters are rich indigo blue, with bright, clear blue on the sides of the body and on the extremities. Sometimes lobsters are pale red, hardly distinguishable from the cooked animal when seen from above. Yet, whatever their color topside, live American lobsters tend to be very lightly pigmented, or even cream-colored, underneath.

Occasionally, fishermen catch American lobsters that are cream-colored above as well as below, but with dark eyes and often with red pigment on the underside of the claws. Or such a cream-colored lobster may have faint traces of blue in its shell, as did one that was exhibited in Boston at the New England Aquarium. Such lobsters are frequently called albinos, although true albinos lack all pigment in eyes and shell. True albino American lobsters apparently have never been taken.

No single factor is responsible for the differences in color of American lobsters. The basic color pattern is inherited, just as are color and texture of hair in man and other mammals. But in an American lobster the actual color that develops may depend partly upon the type and strength of illumination to which the animal is exposed and even more upon its diet.

Thus, Professor F. H. Herrick, who in 1895 published a classic monograph on the American lobster, observed that bluish coloration in this animal can result from prolonged

Fig. 57. Albino American lobster, a rarity in nature. This specimen, a female around nine inches long and judged to be about three years old, was trapped on July 2, 1972, off Pemberton Beach, near Hull, Massachusetts. (Courtesy of New England Aquarium, Mary Price, photographer.)

exposure to sunlight. Recently, John T. Hughes and George C. Matthiessen of the Massachusetts Division of Marine Fisheries reported that lobsters held for a period of years at the Division's lobster hatchery and rearing facility in Oak Bluffs, Martha's Vineyard, and fed primarily quahaugs, clams, scallop viscera, and alewives, turned a deep sky-blue color, which eventually faded into a pale blue–grey. When, however, these bluish lobsters were then fed exclusively on green crabs, they reverted somewhat to the wild-type coloration after the next molt and became identical in coloration with the wild-type following the second molt.

Color in all decapod crustaceans results primarily from the presence of pigments known as carotenoids (after carrots, from which they were first isolated) in the tissues and shell. The major carotenoid of decapod crustaceans is astaxanthin, which is bright red in color. When combined, or conjugated, with protein, the red color of free astaxanthin is replaced by a color characteristic of the particular conjugated protein that is present. For example, in the shell of American lobsters, the most abundant pigment usually is a conjugated protein of astaxanthin that is blue. Eggs of American lobsters contain a green conjugated protein. Green crabs about to molt have a green conjugated protein in the old shell and a brown one in the epidermis and pigmented layers of the new shell.

The reason that diet plays such an important role in development of color in American lobsters and other decapod crustaceans is that carotenoids present in the conjugated proteins of these animals have to be either ingested or produced in the animal's body from ingested carotenoids. These pigments cannot be synthesized from noncarotenoid material, except by plants.

Shrimps, lobsters and crabs turn red when they are cooked because heat breaks down the linkage between astaxanthin and protein, and the astaxanthin is freed. Shrimps, lobsters and crabs that are red before being cooked do not have free astaxanthin, but rather an astaxanthin–protein complex that is red in color.

With regard to coloration, there is an important difference between shrimps, on the one hand, and lobsters, crayfishes and crabs, on the other. This concerns the way in which the colors are manifest. Shrimps have a light, fragile, quite transparent shell, through which the underlying integument is visible. In the integument are numerous pigment-containing cells known as chromatophores. Under the influence of certain hormones that originate within the central nervous system and are released into the hemolymph, the pigments within the chromatophores either concentrate in the center of the cell or migrate to the periphery, as the case may be.

Chromatophores have many branches, and thus a cell in which the pigments are dispersed looks very different from one in which the pigments are assembled into a tiny mass at the center. Furthermore, the area covered by chromatophoral pigments when they are dispersed is much greater than when they are concentrated, so the degree of pigment dispersion largely determines the overall coloration of a shrimp. This may change, rapidly and frequently, in response to changes in illumination and color of background, a fact that explains why common names for shrimps often include some that are descriptive of very different colors (see the Foreword).

In lobsters, crayfishes and most crabs, the shell is thick, strong, and largely opaque, due to pigments that are deposited within the shell. Hence, in these decapod crustaceans, the color of the animal is fairly constant, depending primarily upon the color of pigments within the shell rather than upon the degree of dispersion of pigments within the chromatophores. Only in certain restricted areas is the shell of a lobster, crayfish, or crab more or less transparent, and here the color of the underlying pigments can be seen. In a few crabs, notably the fiddler crab *Uca pugilator* and the ghost, or sand, crab *Ocypode,* the shell is fairly light and semi-transparent, and overall coloration results largely from pigments within the chromatophores.

Sometimes the left half of an American lobster (or of its close relative, the European lobster, *Homarus gammarus*) may be of one color and the right half quite a different color. Professor Herrick and several later investigators described a number of such particolored lobsters: light yellow/bright red; dark green/pale red; blue/white; green–black/light orange; dark green/sky blue; dark blue/light red; dark green/red; white–red/purple–blue.

In one case a bilateral difference in color of American lobster was correlated with a bilateral difference in sex. In 1959, Dr. Fenner A. Chace, Jr., and Dr. George M. Moore described an American lobster that on its left side was orange, with mottling and spots of dark green–brown, and on its right side was similarly mottled and spotted but mostly in shades of blue over a light blue ground color. Externally, the lobster appeared female on the right side and male on the left side. When the lobster was dissected, it was found to have well-developed female reproductive organs on the right side and male reproductive organs on the left. In three earlier reports by other scientists, American lobsters having both male and female reproductive organs were described, but in no case was the bilateral difference in sex associated with a bilateral difference in coloration.

Crabs

We have seen that although crabs appear to be tailless, they have a very small tail, which they keep tucked underneath their body. Due to its small size, this tail and its appendages cannot be used for locomotion. The thoracic legs of a crab are used for walking. In certain crabs, including the blue crab, the last pair of thoracic legs is flattened and paddle-shaped and is used for swimming.

While the tail of shrimp or lobster is among the meatiest and most succulent portions of the animal, the tail of a crab con-

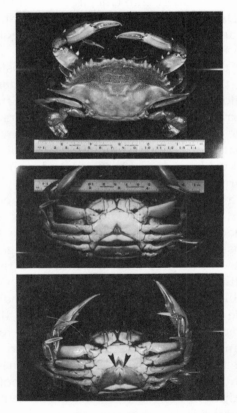

Fig. 58. Female blue crab, as seen from above (top) and below (center). When tail (abdomen) is drawn back and away from body (bottom), genital openings (arrows) are visible on thorax.

tains little meat. The dorsal abdominal muscles are small and very weak, being used solely to extend the tail backward. Virtually the only time at which these muscles are used is during mating, when the abdomen of both male and female must be drawn backward to permit the transfer of sperms.

The ventral abdominal muscles of crabs are somewhat heavier and stronger, particularly in the mature female. While carrying eggs, she uses these muscles to curl her broad, rounded abdomen over the mass of eggs. When not carrying eggs, she uses these same muscles to hold her abdomen tightly in a depression on the ventral surface of her body. Male crabs have a "locking device," consisting of small tubercles on the fifth thoracic segment that secure the triangular or T-shaped abdomen in a depression on the ventral side of the thorax.

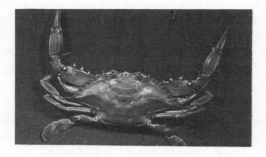

Fig. 59. Male blue crab, as seen from above (top) and below. T-shaped tail (abdomen) is visible (arrow), held tightly against thorax.

Covering both head and thorax of a crab dorsally is a hard carapace. Thus, the boundary between the two body regions is obscure and, as in shrimps and lobsters, one generally speaks of a cephalothorax rather than of the two separate regions. The cervical groove, indicating the boundary between head and thorax, lies just behind the center of the carapace, where it runs generally forward and to each side.

Ventrally, the boundary between head and thorax is well marked, as is the division of the thorax into segments, although only the last five may readily be visible. Also the attachment of the thoracic legs to the exoskeleton is clearly apparent, one pair on each of these last five thoracic segments. The first pair is modified as chelipeds, or claws, while the remaining four pairs are adapted for walking or, in some cases, for walking and the last pair for swimming. In two families of primitive crabs (Dromiidae, Dorippidae), the last pair or last two pairs of thoracic legs are held dorsally, often supporting a piece of sponge or bivalve shell or some other type of sheltering material.

The cephalothorax of a crab is characteristically short and broad and, in some species, greatly extended to the sides. In the blue crab, *Callinectes sapidus*, the paired, widely expanded branchial regions of the carapace terminate in a long, sharp lateral spine. Here the exoskeleton turns sharply inward and downward, to end just above the legs. As a result of anterior–posterior compression and lateral expansion, the branchial chambers are short and wide. Within these chambers the gills are, of necessity, arranged in a broad oval, rather than linearly as in shrimps and lobsters. Indeed, at their base the most anterior pairs of gills in a blue crab "face" forward.

Contrary to widespread popular belief, crabs can walk forward or diagonally, and some species do so quite often. But usually crabs move sideways, particularly when hurrying. The attachment of one pair of chelipeds and four additional pairs of thoracic legs within the short space available at the side of a crab favors sidewise movement over forward movement. When a crab runs sidewise, the legs on the leading side pull the body by flexing, while the legs on the trailing side push the body by extending.

One genus, the semiterrestrial ghost crab, *Ocypode,* can run at great speed. In tests by Dr. Dennis R. Hafeman and Dr. J. I. Hubbard, the species *Ocypode ceratophthalma* ran at an average speed of 1.825 meters per second, or over four miles per hour, on the firm sand of a tidal beach. When on the hard deck of a ship, the crabs ran even faster, the average speed being 2.33 meters per second, or 5.2 miles per hour. These are the highest recorded speeds for any crustacean. During the tests the crabs did not use their last (fifth) pair of thoracic legs or their chelipeds, except for balancing. The second, third, and fourth pairs of thoracic legs did the moving, with the second and fourth legs on the leading side usually being extended first, to be followed by the third leg. On the trailing side, the same sequence occurred, but with a phase lag of about a third of a cycle.

Some observers have reported that when *Ocypode* is running, it does so with one side leading for a while. The crab stops abruptly, rotates its body, and then runs with its other side leading. The process of rotation is repeated. In this way, the flexor and extensor muscles of the legs on each side are alternately used and rested.

In the anterior portion of the cephalothorax of a crab are the mouth parts, grouped around the opening to the esophagus. These mouth parts are generally similar to those of shrimps and lobsters. The outermost pair is the third maxillipeds, used for holding food. Under and in front of these are two more pairs of maxillipeds and two pairs of maxillae, also used for holding food, and a pair of mandibles, or jaws, which push the food into the esophagus.

The foregut of crabs, like that of lobsters, has in its walls many ossicles, or small hard plates and projections, that articulate with one another in a complicated way and serve as a place of attachment for muscles that move the foregut. According to American biologists Robert Pyle and Eugene Cronin, the blue crab has in or associated with its foregut at least 50 ossicles and over 80 muscles. These effect a churning action of the foregut and a grinding of the gastric mill that break down particles of food that have been swallowed. The gastric mill of crabs resembles that of lobsters in consisting of one dorsal and two lateral teeth situated at the constriction that separates the large anterior chamber of the foregut from the smaller posterior chamber.

The midgut originates approximately where ducts from the midgut glands enter the posterior chamber. Behind this chamber, the midgut appears as a small tube, scarcely three-eighths of an inch in length in a full-grown blue crab.

The midgut glands consist of three pairs of lobes, one pair extending forward and to the sides, a second pair extending laterally toward or over the gills, and a third pair leading back toward and, in some species, into the abdomen. The midgut glands may fill much of the body cavity, although their extent at any one time depends largely upon their content of food reserves and water.

As in shrimps and lobsters, digestion of food in crabs takes place partly in the anterior chamber of the foregut, partly in the posterior chamber, and partly within tubules of the midgut glands. A bristly filter in the ventral wall of the posterior chamber prevents all but the most finely divided material from passing up the ducts into tubules of the midgut glands. Since the lumen of the midgut glands is continuous with that of the midgut, these glands are diverticula of the midgut.

In the blue crab, another diverticulum arises from the midgut just behind the posterior chamber. A pair of tubes, known as midgut ceca, runs from the dorsolateral surface of the midgut forward and laterally, ending in coils that lie just above the first large lobe of the midgut glands. These ceca are translucent and difficult to see in dissection. Lining the lumen of the midgut ceca are cells like those lining the midgut. Both groups of cells probably function in the absorption of food.

The hindgut makes up the remainder of the digestive tract. It runs between the lobes of the midgut glands, under the heart, and into the abdomen, where it follows a straight course to its posterior opening, the anus. Only a slight swelling is present in the most posterior portion of the hindgut, hardly enough to justify calling this region a rectum. The entire hindgut is lined with chitin.

In the second or third abdominal segment, the hindgut of the blue crab gives rise to a cecum. From its origin on the left

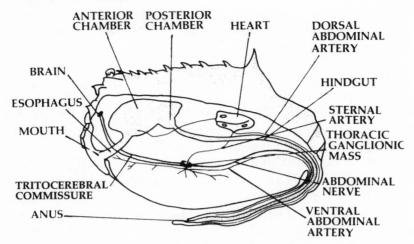

ANTERIOR POSTERIOR
CHAMBER CHAMBER   HEART   DORSAL
                         ABDOMINAL
                         ARTERY

BRAIN

ESOPHAGUS   HINDGUT

MOUTH   STERNAL
        ARTERY
        THORACIC
        GANGLIONIC
        MASS

TRITOCEREBRAL   ABDOMINAL
COMMISSURE   NERVE

ANUS   VENTRAL
       ABDOMINAL
       ARTERY

Fig. 60. Internal anatomy of blue crab, showing major organs of digestive, circulatory and nervous systems. Dorso-lateral view. Midgut glands, which extend through much of body, omitted here.

side, the cecum runs forward and over the hindgut, terminating in closely packed coils on the right side. The function of this cecum is not clear, but this organ may be involved in the regulation of salts in the hemolymph when a crab is exposed to dilute media.

In a crab, circulation of blood takes place much as in a shrimp or lobster. Arteries carry blood dorsally from the heart forward into the head and viscera and backward into the abdomen. A sternal artery carries blood ventrally, where main branches direct it both forward and back. Further branching of main arterial vessels leads the blood into thin-walled capillaries, where exchange of gases and foodstuffs between blood and tissues can occur. The blood collects in venous sinuses, goes to the gills, and then enters the pericardial sinus surrounding the heart. Here, when the heart relaxes and its three pairs of ostia open, the blood enters the heart. In keeping with the breadth of a crab's body, the heart is broad, filling much of the pericardial sinus, or pericardium. In the small tail of a crab, the ventral abdominal artery is of relatively small size compared with the same artery in the muscular tail of a lobster.

The nervous system of all crabs, except the most primitive, has undergone a high degree of fusion. All ventral ganglia are fused into a single thoracic ganglionic mass, which lies near the floor of the cephalothorax and through which the sternal artery descends. From the periphery of the thoracic ganglionic mass, nerves radiate out to the appendages all the way from the mandibles to the last thoracic legs. An abdominal nerve emerges posteriorly at the midline and supplies the muscles and appendages of the tail.

Connecting the thoracic ganglionic mass with the brain, or supraesophageal ganglion, are the two long, large nerves that pass on either side of the esophagus and are known as the circumesophageal connectives. Slight swellings on the connectives mark the position of the stomatogastric, or connective, ganglia that supply nerves to the foregut. The tritocerebral commissure links the two connectives in crabs, as in shrimps and lobsters.

The kidneys of crabs lie on the interior ventral surface of the

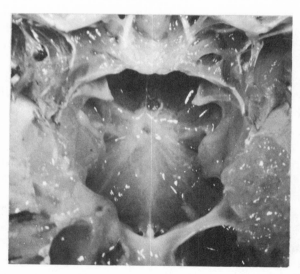

Fig. 61. Thoracic ganglionic mass of blue crab, lying near floor of cephalothorax and with nerves radiating to mouth parts, thoracic legs, and tail. Sternal artery descends through hole visible in thoracic ganglionic mass.

body, just posterior to a position between each antenna and the corresponding eyestalk on the same side. Due to their color, which is pale green, yellow, or green–brown, the kidneys are also called green glands; due to their position on the second, or antennal, segment, they are also often called antennal glands.

In structure, the antennal glands of crabs are similar to those of lobsters. There is a glandular portion, which secretes urine and regulates salts, and a large, many-lobed, thin-walled bladder, in which urine is temporarily stored. The main lobe of the bladder lies above the glandular portion of each kidney, but the remaining lobes extend out in several directions. Because of the delicacy of the lobes, it is almost impossible to see them unless they are fixed in alcohol or injected with India ink or a powdered dye, such as carmine. Urine passes from the glandular portion of each antennal gland upward into the bladder and then to the exterior via a duct. The opening of the duct, which lies at the base of the antenna, is covered by a calcified, movable cover, called an operculum.

The gills of crabs differ from those of lobsters, where each gill consists of many filaments arranged, plume-like, around a central axis. In crabs two rows of closely set, leaf-like plates, or lamellae, are attached to the central axis of all or, in some species such as the blue crab, all but one pair of gills. In the blue crab, one anterior pair of gills has only one row of lamellae. Gills of crabs are known as phyllobranchs, after the Greek words *phyllon*, meaning leaf, and *branch*, meaning gill. There are eight gills on each side of a blue crab's body.

Water enters the branchial chambers of crabs primarily through an anterior opening above the base of each claw, or cheliped, and to a much less extent through openings at the base of the other thoracic legs. In the blue crab, when the chelipeds are raised and held forward, the opening at the base of the chelipeds is very large and nearly circular. When the chelipeds are folded against the body, the opening is a wide slit, which becomes narrower when the third maxillipeds are brought close to the midline, for a flange at the base of each third maxilliped reduces the width of the slit. Bristle-like setae arising from the basal portion of the cheliped filter some of the water entering the slit.

In crabs, the opening at the base of the last pair of thoracic legs may or may not be important in the entry of water. According to Drs. Arudpragasam and Naylor, who studied pathways of gill ventilation in several species of crabs, the more flattened and shortened the body of a crab and therefore the more it diverges from the elongated body and laterally facing gills of a lobster or shrimp, the more important in the entry of water are the anterior openings at the base of the chelipeds and the less important are posterior openings.

As in shrimps and lobsters, the current of water through the branchial chambers of crabs is maintained by the beating of the gill bailer, which lies in the channel at the anterior exhalant opening of each branchial chamber. After entering through the openings at the base of the chelipeds and, to a less extent at the base of the thoracic legs, the water passes under the gills, up between the gills, over the gills, and out through the exhalant aperture. Periodically, as a result of reversal in the action of the gill bailer, the direction of the respiratory current is reversed. This aids in cleaning the gills of debris and tends to divert water back over the gills that lie in the posterior part of the branchial chambers.

The ovaries of a female blue crab are connected to each other just behind the foregut and extend forward and backward through the body. Thus, in blue crabs, as in American lobsters, the ovaries appear roughly H-shaped. In early stages of its development, each ovary of a blue crab is thin and white, with a short lateral arm. It still appears this way immediately after the female has shed her shell and, as a soft crab, has mated. But ovarian growth starts soon thereafter and, several months later, results in a very large ovary, which is orange because the eggs are full of orange yolk. Each ovary now may extend far out to the side of the body and into the first abdominal segment.

From the ovaries, paired oviducts run forward and downward for a short distance, then widen to form an oval-shaped structure known as a seminal receptacle. Here are stored the sperms that the female blue crab receives from the male blue crab during mating. Each seminal receptacle slants backward and downward and then narrows into a short tubular vagina,

OVARY   FOREGUT   MIDGUT
(LATERAL LOBE) (ANTERIOR CHAMBER)   HEART   GLAND   GILLS

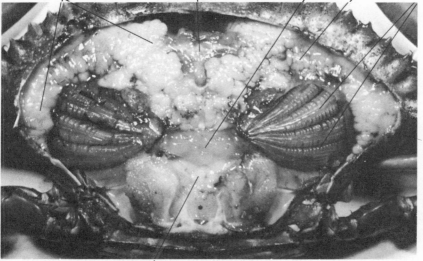

OVARY
(POSTERIOR LOBE)

Fig. 62. Dorsal view of internal organs of a female blue crab, with
well-developed orange, yolk-filled ovaries extending far out to
sides of body and back to abdomen.

which runs ventrally to an opening on the sixth thoracic
segment. Although the oviducts and the dorsal portion of the
seminal receptacles are soft and unlined, the ventral portion
of the seminal receptacles and the vagina are hard, being lined
with chitin. At the time of ecdysis, this lining is shed, along
with other portions of the exoskeleton.

In an immature blue crab the seminal receptacles are small
and white. Yet, in a mature crab immediately after copulation,
the seminal receptacles are enormously distended, at times
equal in size to the heart; and they are pink in color, due to the
presence of a gelatinous "sperm plug" that keeps the sperms
secured within the receptacles. Later, after the sperm plug
has been absorbed, the receptacles are again white.

At the time of ovulation, when ripe eggs are released from
the ovaries and move down the oviducts, sperms fertilize the
eggs either within the oviducts or within the seminal receptacles. When fertilized eggs emerge from the vagina, they be-

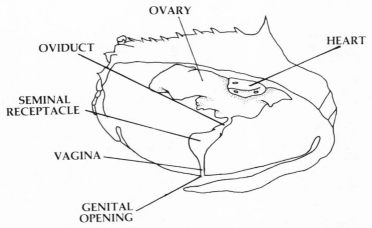

OVARY

OVIDUCT

HEART

SEMINAL
RECEPTACLE

VAGINA

GENITAL
OPENING

Fig. 63. Female reproductive system of blue crab, dorso-lateral view. Genital tract of one side only is shown.

come attached to the pleopods of the female and remain there until ready to hatch into the first larval stage. Yet many sperms remain within the seminal receptacles and many eggs within the ovaries, so usually a second "laying"occurs, after which the ovaries appear collapsed and grey or brown in color as they begin to degenerate. Yet even now, enough sperms remain within the seminal receptacles to fertilize several more batches of eggs, were the eggs able to ripen.

In the male blue crab the testes consist of a pair of slender, convoluted, opaquely white arms lying on the dorsal surface of the midgut glands. Medially, the terminal portion of each arm passes around the posterior end of the foregut and joins with the other arm to form a short cross-bar. Just anterior to the cross-bar a tiny tube, the vas efferens, connects each arm of the testes with a much-coiled vas deferens. The vas efferens is difficult to find since it is concealed within the testis and the coils of the vas deferens.

The vas deferens consists of several portions. The first, known as the anterior vas deferens, is white and tightly coiled and lies close to the middorsal line between the foregut and the heart. Here the sperms are gathered in egg-shaped bundles, called spermatophores, and stored. In the second por-

FOREGUT   MIDGUT
(ANTERIOR CHAMBER)   TESTIS   GLAND

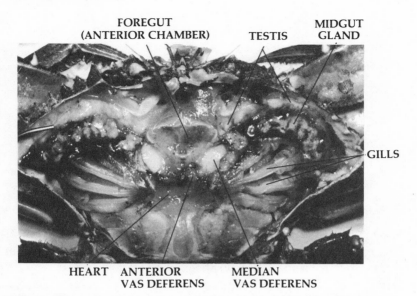

GILLS

HEART   ANTERIOR   MEDIAN
         VAS DEFERENS   VAS DEFERENS

Fig. 64. Internal organs of a male blue crab, with slender, convoluted arms of testes lying on dorsal surface of midgut glands and anterior and medial vas deferens visible between foregut and heart.

tion, the median vas deferens, the coils form a large mass and appear pebbled pink, due to their content of material that subsequently is deposited in the seminal receptacles of the female during copulation and forms a sperm plug.

The third portion, the posterior vas deferens, is long, convoluted, greenishly translucent, and almost empty except during passage of the spermatophores. The final portion, the penis, is a short, translucent tube at the base of the last pair of thoracic legs. The penis lies permanently within a groove in the first pair of abdominal appendages, the copulatory pleopods. These, in turn, are inserted into the seminal receptacles of the female during copulation. Also fitted into the groove on the copulatory appendages of the male is his second pair of pleopods, which during copulation act as pistons to push the spermatophores along the groove, where they break up and release the sperms.

The T-shaped abdomen and elongated, grooved copula-

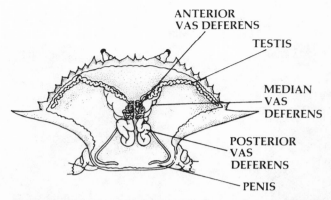

Fig. 65. Dorsal view of male reproductive system of blue crab.

tory pleopods of the male blue crab are his most distinguishing external sexual characteristics. In contrast, the abdomen
of the mature female blue crab is broad and rounded, and her
pleopods are relatively short and fringed with hair-like setae,
to which eggs are attached during development. Coloration
also serves to separate the sexes. Normal coloration of both
male and female blue crabs consists of a dark blue–green or
gray–green carapace and bright blue and blue–green legs,
with scarlet markings. Except for the appendages and the
female abdomen, the underparts are white. In the male, the
greater portion of the chelipeds, or claws, is blue–or
gray–green, with dull purple "fingers," whereas in the
female there is more blue on the chelipeds and the "fingers"
are bright red.

It may interest the reader to learn that just as there are blue
American lobsters, so also there are blue blue crabs. Some
years ago a specimen of blue crab was described as having a
carapace of robin's egg blue and appendages of pale blue with
traces of pale red. The under surface of the body was white.
Also, just as parti-colored American lobsters exist, so do
parti-colored blue crabs. One such specimen was described as
being gray on the left side and brownish on the right.

A tendency toward albinism occurs in blue crabs, as it does
in American lobsters. In a specimen of blue crab captured in
Tampa Bay in 1965, the carapace was normal in coloration.
Except for spots of color, however, the chelipeds and most of
the thoracic legs were white.

Five

Mating and Spawning

At the time of mating, shrimps, lobsters and crabs perform certain rituals, during which the male transfers to the female one or more packets of sperms known as spermatophores. The spermatophores remain with the female until eggs within the ovaries of the female ripen and are ready to descend the oviducts and be fertilized by the sperms. If the male has placed the spermatophores within the female's oviducts, fertilization takes place internally. But if the male has placed the spermatophores on the external surface of the female, fertilization occurs externally. Whether fertilized or unfertilized, the emission of eggs from the genital pores of the female is known as spawning. Following spawning, the female usually carries fertilized eggs about with her as they undergo development. Hatching occurs when egg development is complete, and a tiny free-swimming larva emerges from each egg case.

Although this is the general pattern of mating and spawning in decapod crustaceans, the details of these activities vary among the species. This will be evident as one reads this chapter. It should be pointed out that the examples described here are necessarily limited by the des-

criptions available within the literature, for many species of
shrimps, lobsters and crabs have never been seen to mate or
spawn. In nature, mating often takes place only when the
animals are in seclusion. In the laboratory it may never take
place.

Shrimps

Among the more primitive shrimps, the penaeids, there is
one known as *Sicyonia carinata* that lives in the Gulf of Naples.
An Italian biologist, Dr. Arturo Palombi, discovered that a
form of courtship takes place in this species. A male shrimp
approaches a female and rubs his spearlike, forward-
projecting process known as the rostrum back and forth along
the ventral surface of the female's thorax and abdomen. In
response, she lifts her abdomen and the male slides beneath
her. Positioning himself perpendicular to her, he grasps her
with his thoracic legs. During the next 15–20 seconds he
transfers a spermatophore to the area on her ventral surface
known as the thelycum (see Chapter 4), and then he relaxes
his grip on her and moves away. Ripe eggs emerge from her
genital openings several days later and are fertilized by
sperms from the spermatophore. Her eggs, like those of other
penaeid shrimps, disperse within the water and are heavy
enough to sink to the bottom and remain there during
development.

In the mating of another penaeid shrimp, *Penaeus japonicus,*
according to a Japanese biologist, Dr. Motosaku Hudinaga, a
hard-shelled male selects a female that is about to shed her
shell and follows her as she walks around. After she sheds,
the male approaches and grasps her. With their ventral sides
aligned and in contact, the hard male and the soft female
swim together for several minutes, while the male transfers a
spermatophore to the female. Then the male and the female
separate. The female may not spawn until she has shed her
shell and re-mated several times. In spawning, she swims
slowly with the tip of her abdomen bent slightly downward
and her thoracic legs held tightly against her body, her
pleopods, or swimmerets, moving vigorously forward and

backward. When eggs emerge from her genital openings, they are fertilized by sperms from the spermatophore. The movements of her pleopods cause the fertilized eggs to be scattered behind her in two streams. The eggs sink to the bottom and remain there while they develop. The female requires only three or four minutes to lay up to 700,000 eggs.

The white shrimp, *Penaeus setiferus*, of the South Atlantic and Gulf coasts, can mate and often does when the female is hard. As in *Penaeus japonicus*, spawning in the white shrimp may not occur for a time after the female has mated. Meanwhile, she may shed the shell to which the spermatophore is attached and may again mate and receive a new spermatophore. Spawning begins in March and April and may continue at intervals throughout the summer. In a single spawning, from 300,000 to 1,000,000 eggs emerge from the genital openings of a female and are fertilized by sperms from a spermatophore on her thelycum. The fertilized eggs are dispersed through the water and sink to the bottom, where they develop rapidly and may hatch within 24 hours.

Caridean shrimps mate and spawn somewhat differently from penaeid shrimps. Two English biologists, Drs. A. J. Lloyd and C. M. Yonge, have described the process in *Crangon crangon*. A male approaches a female that has just shed her shell, turns her over on her back, and bends his body U-shaped over her soft body at the junction of her thorax and abdomen. He does not use his appendages to grasp her, but maintains his position above her by muscular effort. Next the male shrimp applies two strand-like spermatophores to the ventral side of the female, close to the openings of her genital ducts. Male and female remain in contact for about five seconds, after which they fall apart and are motionless. The female quickly revives and buries herself in the sand. The male soon swims away, ready to mate with another female.

Within 24 hours of mating, the impregnated female shrimp seeks a secluded spot, where she lies on her side with her abdomen bent under her thorax. Eggs begin to emerge from her genital openings and move posteriorly in chains, their progress aided by movement of the first pair of pleopods. When the eggs reach the spermatophore, they are fertilized by the sperms embedded there. The newly fertilized eggs come to rest in a "brood pouch" formed by the flexed abdomen of the female, her pleopods, and her tail fan. The female secretes a cementing substance that hardens and secures the eggs to the hairs of the pleopods. Here they remain during incubation.

French scientists Drs. Henri and Louise Nouvel found that the caridean shrimps *Athanas nitescens, Palaemon squilla,* and *Alpheus dentipes* also mate a few hours after the female has shed her shell and while she is still soft. A hard male may approach a female from behind or from the side, pass part way under her, and halt perpendicular to her, his ventral surface uppermost and his genital openings facing her first abdominal segment. The male quite quickly transfers the spermatophores to the female. She, in turn, may spawn immediately thereafter or within several hours. She carries her fertilized eggs for two or four weeks before they are ready to

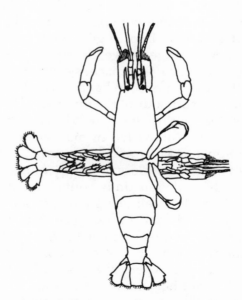

Fig. 66. Mating in the caridean shrimp *Athanas nitescens,* with female above and male perpendicular to her. (Redrawn from H. and L. Nouvel)

hatch. In cold weather though, her eggs may require several months to reach the hatching stage.

Lobsters

A female American lobster, *Homarus americanus,* can mate only during the 24 to 48 hours following her ecdysis, or shedding of the shell, and while she is still soft. The male will have shed his shell several weeks before and by the time of mating is fairly hard. According to W. Templeman, who in 1933 published a rather complete account of mating in the American lobster, a male lobster approaches a soft female and with his appendages turns her over on her back. There she remains motionless, her large claws lying straight out in front. Only her pleopods remain in motion. The male moves over her and stretches his claws forward, often resting them on the motionless claws of the female. With his thoracic legs he grasps her body. Both male and female lobsters keep their abdomen extended, their pleopods moving back and forth vigorously.

The male American lobster now inserts his first pair of pleopods, which at maturity became modified as copulatory appendages, into the seminal receptacle on the ventral side of the female's thorax between the third and fifth pairs of thoracic legs. Being long, hard, smooth, and tapering, and with a groove on their inner surface, the male's copulatory pleopods readily guide the elongated gelatinous spermatophores from his genital openings at the base of his fifth pair of thoracic legs into the seminal receptacle of the female.

Mating is completed within 30 to 110 seconds, the average duration being about one minute. After mating, a soft gelatinous substance can be seen at the opening of the seminal receptacle of the female. This gelatinous material gradually hardens, and within nine or 10 hours of mating it has plugged the opening completely, making further mating impossible. Several days later, when the gelatinous capsule surrounding the sperms in the spermatophores has hardened, the plug at the opening of the seminal receptacle drops off.

Until the gelatinous plug that forms during mating has hardened, there is nothing to prevent a soft female from mating again with another hard male lobster. During certain experiments, Dr. Templeman found that nine different hard-shelled males in succession were able to mate with one soft-shelled female within about one and one-half hours. He also found that a male lobster, following one mating, was ready for a second mating with another female several hours later. Usually a male lobster is slightly larger than a female with which it mates. Male and female lobsters that differ greatly in size cannot mate successfully.

Anywhere from one month to two years after mating, a female American lobster spawns. She lies on her back and props up the anterior part of her body with her large claws, which she stretches forward and somewhat to the sides. The eggs emerge through her genital openings at the base of the third pair of thoracic legs. They flow to the rear, six to eight abreast, past the opening of her seminal receptacle, where they are fertilized by sperms from the spermatophores. Then the eggs enter a pocket formed by the upward curve of her tail and become cemented to the hairs on her pleopods. Egg-laying usually requires several hours to be completed and generally occurs once every other year, between early June and September.

The number of eggs that are laid at one spawning varies with the size of the lobster. A seven-inch female American lobster lays approximately 3,000 eggs, while an eighteen-inch female lays around 75,000 eggs. When laid, the eggs are dark green, almost black, and only about 1/16 inch in diameter. Gradually turning brownish in coloration, the eggs are carried by the female for about a year before they hatch.

It has long been thought that the eggs of the South African rock lobster, *Jasus lalandii,* are fertilized within the female, although no one seems to know just how this can be effected, for the first pair of pleopods of the male is not modified as copulatory appendages. According to biologist Cecil von Bonde, who in 1936 published an extensive account of reproduction in the South African rock lobster, mating takes place about two hours after the female has shed her shell. A male rock lobster approaches a soft female, turns her over on her back, and applies his ventral surface to hers in a head-to-head position. Almost immediately spermatophores emerge from his genital openings. Each spermatophore, which consists of clumps of sperms embedded within a fluid matrix, is sticky and jelly-like. Dr. von Bonde believed that a spermatophore is introduced into each oviduct of the female, that the matrix disintegrates, and that the sperms are freed, ready to fertilize the ripe eggs as they descend. Mating is completed within 30 to 60 seconds.

About two or three days later, the female rock lobster lays between 3,000 and 20,000 eggs, depending upon her size. Standing on her thoracic legs, she flexes her abdomen under her cephalothorax and holds her pleopods close together, so that they form a small tunnel that is closed at its end by the tail fan. The fertilized eggs come out of her genital openings in a continuous stream and move tailward through the tunnel, their progress aided by a posteriorly directed current of water created by the beating of her pleopods. The first eggs to be laid become cemented to the first pair of pleopods, and the remaining eggs to successive pairs. It takes from three to four hours for egg-laying to be completed.

In the Indo-West Pacific spiny lobster *Panulirus penicillatus,* the male during mating places two spermatophores externally on the ventral surface of the female's thorax, just behind the opening of her oviducts. Each spermatophore consists of a strand of sperms in a granular capsule embedded within a noncellular matrix. When first formed, the spermatophoric mass, made up of the two spermatophores, is light in color and pliable like putty, but it gradually hardens and blackens, probably as an adaptation for life in shallow, turbulent water. Subsequently, when eggs emerge from the female's oviducts, she uses a small claw at the tip of her fifth thoracic legs to break open the hard spermatophoric mass and release the sperms within it. The eggs, now fertilized externally by these sperms, become cemented to hairs on the pleopods of the female and develop there.

In three other species of *Panulirus,* the female spiny lobster receives from the male pliable, putty-like spermatophores that gradually harden and blacken as she carries them externally on her thorax. In these species, fertilization is external. The same is true of a southeast African spiny lobster, *Palinurus delagoae*, which lives in deep water and has a soft, sticky, unprotected spermatophoric mass.

Anomuran Crabs

From Chapter 1 it may be recalled that anomuran crabs are members of the infraorder Anomura, in which the last pair of thoracic legs may be either hidden within the branchial chambers or exposed but so small as to be easily overlooked. Thus, anomuran crabs may appear to have only four pairs of thoracic appendages; that is, one pair of claws and three pairs of thoracic legs. Also included in the infraorder Anomura are the hermit crabs, which have a long, soft, spirally twisted

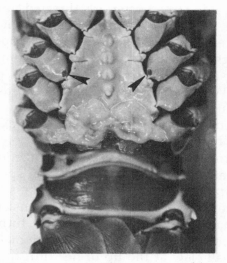

Fig. 67. Soft, sticky, unprotected, opaque-white spermatophoric mass of deep-water southeast African spiny lobster *Palinurus delagoae*. Mass adheres to underside of thorax of female at base of fifth pair of thoracic legs, with genital openings (arrows) visible at base of third pair of thoracic legs. First pair of female's leaf-like abdominal appendages (pleopods) are visible at base of photograph. (Courtesy of P. F. Berry and *Crustaceana*)

abdomen that fits within the chamber of a discarded snail shell.

An Indian biologist, Dr. S. Kamalaveni, was able to observe a hermit crab, *Clibanarius olivaceous,* during mating. To do this, she removed the snail shell from each of several male and female hermit crabs and placed the denuded animals in an aquarium. Dr. Kamalaveni found that a male hermit crab is attracted to a female hermit crab that has just finished shedding. The male seizes the soft female and holds her with his claws, intertwining his thoracic legs with hers. This brings the ventral surface of the thorax of male and female close together, with the genital openings of the two almost opposite each other.

When the two hermit crabs have been in this position for about 20 minutes, the spermatophores of the male begin to emerge from his genital openings as two long ribbons of sticky, sperm-containing mucus, which quickly adheres to the thorax of the female, close to her genital openings. On contact with the brackish water in which these hermit crabs usually are found, the spermatophores apparently explode. The sperms thus released from the spermatophores pass through the genital openings of the female into her oviducts, where they fertilize the eggs that are descending the oviducts. The male and female hermit crabs remain, each with its thorax pressed against the thorax of the other, for nearly an hour and a half. In this way most, if not all, of the sperms that are released from the spermatophores are kept close to the genital openings of the female, and most of the descending eggs are fertilized. About 200 brick-red eggs are laid. By means of a cementing substance that the female secretes, the eggs become attached to the first three abdominal appendages of the female on her left side, but not to the fourth abdominal appendage, which is rudimentary. In a female hermit crab abdominal appendages are absent from the right side. Hatching of the eggs occurs about 12 days after spawning.

Probably the best known anomuran crab is the king crab *Paralithodes camtschatica,* which becomes sexually mature when it is five to six years of age and when its carapace measures three and a half to four inches in width. Mating of king crabs takes place in fairly shallow water either by day or by night, but only from late March to late May. The size of the mating partners seems unimportant, for mature male king crabs mate with mature females that are much smaller or much larger than the males. Female king crabs do not have a seminal receptacle. Therefore, the male king crab must be present during the female's ovulation, if fertilization of her eggs is to be accomplished.

Premating behavior in king crabs begins when a hard-shelled male approaches a female that is close to ecdysis and grasps her in a face-to-face position that has been termed "hand-shaking." The male may carry the female about with him in this way for up to seven days. When the time for her ecdysis has arrived, the male may assist her in shedding her shell. Then he grasps her and turns her over so that her ventral surface and his lie close together. Next the male withdraws his small fifth pair of legs from under the carapace and doubles them back until their broad hairy tips almost touch his genital pores at their base.

Fig. 68. "Hand-shaking" by king crabs, a form of premating behavior, in which hard-shelled male (above) carries female about with him until she sheds shell and mating can ensue. [© National Geographic Society, Robert Sisson, photographer]

Now strand-like spermatophores begin to emerge from the genital pores of the male. According to some observers, he extends his fifth pair of legs and moves the tips vigorously amongst the exposed pleopods of the female, thus spreading the spermatic material over her pleopods. She assists by beating her pleopods rapidly and working the material firmly in amongst the hairs on the pleopods. Masses of spermatic material also lodge at the base of her third pair of thoracic legs, where her genital pores are situated.

At about this time eggs start to descend the oviducts of the female. As they emerge from her genital openings, they collect in the genital pouch that is formed by her partially flexed abdomen. Here the eggs are fertilized by sperms that are already present. The female then holds her abdomen tightly against her body until the fertilized eggs become securely attached to her pleopods, a matter of two or three days. A female king crab lays from 150,000 to 400,000 eggs at a single spawning, and she incubates them for almost a year before they hatch.

True Crabs

Several means have been described on previous pages whereby a male shrimp or lobster or hermit crab can transfer spermatophores to a female and his spermatophores are secured on the female until time for her eggs to be fertilized. Chance, nonetheless, plays a role. For a spermatophore that is attached externally may be dislodged—or lost along with a cast shell at ecdysis. And sperms such as those of some hermit crabs that have to find their way from the exterior into and up the oviducts of the female may not get there.

Except for two groups of primitive forms, the true crabs, constituting the infraorder Brachyura, have developed an apparently reliable means of transferring and storing spermatophores. The basal portion of the male's first pleopods has a long, hollow, spine-like process that serves as an intromittent organ. His second pleopods have a short, solid, spine-like process that, during copulation, pushes the spermatophore through the hollow portion of the corresponding first pleopod. The spermatophore enters the hollow first pleopod via the penis, which is the short, slender terminal portion of the vas deferens that in many species is kept permanently within an opening at the base of the first pleopod. During copulation, the male crab inserts his first pleopods into the genital openings of the female and places his spermatophores directly into her seminal receptacles.

In the blue crab mating has been observed many times. A hard male blue crab finds a female that is soon to shed her shell and that still has the triangular abdomen characteristic of immaturity. Hooking his second thoracic legs on her body between her second and first thoracic legs (claws), he carries her with her dorsal side up for two days or more. During this time, fishermen may refer to the pair as a "buck-and-rider" or a "hard doubler." The female then sheds her shell, while the male hovers over her. The female, now very soft, turns over on her back and unfolds her abdomen, exposing her genital pores. The male again seizes her, this time with her ventral side uppermost. Mating takes place and may go on for as long as 12 hours. The pair at this time is often called a "soft doubler." After mating, the male continues to carry the female beneath him with her dorsal side up for another two days or more.

Fig. 69. Pre-mating "dance" of male blue crab, *Callinectes sapidus*. Female is visible in background. (Courtesy of the Virginia Institute of Marine Science, Gloucester Point, Virginia)

Fig. 70. Female blue crab sheds shell while male hovers over her. (Courtesy of the Virginia Institute of Marine Science, Gloucester Point, Virginia)

Fig. 71. "Soft doublers": male blue crab (above) carries soft female in mating position, with her ventral side up, underneath male. (Courtesy of the Virginia Institute of Marine Science, Gloucester Point, Virginia)

The female blue crab mates only once and only immediately after the ecdysis at which she acquires the broad, rounded abdomen of maturity. By attending the female before and during this molt, the male is sure to be present when the female is ready to mate. By carrying her during and after mating, he protects her while her shell partially hardens.

A female blue crab spawns from two to nine months after mating. Ripe eggs leave her ovaries and enter her seminal receptacles, where they are fertilized by sperms placed there by the male during mating. The fertilized eggs descend the genital ducts of the female, emerge from her genital openings, and are cemented to the hairs of her pleopods. In a single egg mass, called a "sponge," a female blue crab may lay about two million eggs, each about one one-hundredth of an inch in diameter. She may require two hours to complete the task. Yet in her seminal receptacles she may retain enough sperms from her single mating to fertilize another batch of eggs, and she often has a second sponge later in a summer.

According to biologists C. Dale Show and John R. Neilson, for about seven days before a mature female Dungeness crab, *Cancer magister*, sheds her shell, a hard, mature male crab of that species carries the female about with him. During this time he keeps the female beneath him, her ventral surface uppermost and in close contact with his ventral surface. On about the eighth day, the female seeks to escape from the male, and he eventually permits her to turn over, so that her dorsal surface now lies next to his ventral surface. With his claws and legs the male encircles the female as she begins to withdraw from her old shell. When the shell has been shed,

Fig. 72. Frontal view of Dungeness crabs mating, with male above and female below, her ventral side uppermost. (Courtesy of C. Dale Snow)

Fig. 73. Posterior view of Dungeness crabs mating, showing position of abdominal flaps. Abdomen of female, below, is outside abdomen of male, above. (Courtesy of C. Dale Snow)

the male shoves it away with his claws. About an hour and a half later, the male turns the now soft female over so that she again lies ventral side up. The female then extends her abdomen, and the male inserts his copulatory pleopods into her seminal receptacles. The mating crabs remain in copulatory position for 30 minutes to two hours, after which the male may carry the female around in a postmating embrace for a couple of days before releasing her. Within a few months of mating, the female Dungeness crab lays up to one and a half-million whitish eggs, which become cemented to her pleopods. As the eggs develop, their color gradually changes to salmon-pink and then to red.

The male European edible crab, *Cancer pagurus*, also courts the female before mating, according to Dr. E. Edwards, who has observed and photographed this process. From several days to several weeks before a female sheds her shell, a hard male may continuously attend her, maintaining a position astride her back. When the female finally sheds, the male remains astride her, although he supports his own weight on his thoracic legs. Sometimes the male assists the female by pushing her carapace off with his claws. After shedding, the female lies motionless on the bottom. The male gently turns

Fig. 74. Male European edible crab assisting female to shed her shell. (Courtesy of Eric Edwards and Ministry of Agriculture, Fisheries, and Food, London)

Fig. 75. European edible crabs mating. Soft-shelled female lies underneath hard-shelled male, her cast-off shell visible at left. (Courtesy of Eric Edwards and Ministry of Agriculture, Fisheries, and Food, London)

Fig. 76. Female European edible crab, with abdomen (top) pulled back to reveal sperm plugs (arrows) in genital openings on ventral surface of thorax. (Courtesy of Eric Edwards and Ministry of Agriculture, Fisheries, and Food, London; also *Crustaceana*)

Fig. 77. Sperm plugs removed from genital tract of female European edible crab. Coin is one inch in diameter. (Courtesy of Eric Edwards and Ministry of Agriculture, Fisheries, and Food, London; also *Crustaceana*)

her on her back and, with his claws, unfolds her abdomen, thus exposing her genital openings. He then inserts his copulatory pleopods. After two or three hours of mating, the male withdraws his copulatory pleopods, leaving two white sperm plugs that prevent the loss of sperms from the seminal receptacles and the entry of sea water. The male may still attend the female for several days. Some weeks after mating, the sperm plugs will have disappeared externally, although portions of them will remain internally for many months.

In these three examples of premating and mating behavior, the crabs are aquatic. Thus, they may safety mate while the female is soft, and it is probably advantageous for them to do so, since a male crab can insert his copulatory pleopods more easily when the female genital openings and the lower portion of the female genital tract are soft and pliable. Nonetheless, some aquatic crabs, notably certain spider crabs (Majidae), pea crabs (Pinnotheridae), and pebble crabs (Xanthidae), may mate while the female is hard.

The burrowing crab, *Corystes cassivelaunus*, so-called because it lives submerged and usually buried in sand, also mates while the female is hard. In this species, each genital pore of the female is normally closed by a hard, calcified cover known as an operculum. For a short time while the female remains hard, that is, during the intermolt period, the operculum becomes flexible and translucent and can be pushed inward.

Dr. R. G. Hartnoll has observed mating of the burrowing crab in the laboratory. A hard-shelled male crab selects a hard-shelled female crab having soft flexible genital opercula and, with his long claws, holds her with her carapace against his undersurface. If disturbed, he lifts the female somewhat aloft and away from his body by holding her in one claw and then runs along the bottom. After several days, the male and female "face" each other, with the male now grasping the female with his second and third thoracic legs and with his claws now resting on her carapace. The male uses his fourth and fifth thoracic legs to burrow both crabs in the sand. Male and female extend their abdomen, the female's overlapping the male's, and the male inserts his copulatory pleopods into the female's soft genital orifices. During mating, the pair remains partly or completely buried. After mating is completed, the male releases the female. She spawns several days later. The female's genital opercula then become hardened and calcified once more.

It is of interest that the snow crab, *Chionoecetes opilio,* also goes through a preliminary courtship, during which the male crab holds the smaller female crab aloft with one claw. Although the female crab may resist, she is not able to escape from the male until he finally lowers her to the substratum. In this type of mating, which has been observed by Dr. J. Watson, the female has recently shed her shell.

When true crabs mate on land, the hardness of the female becomes more important, for a copulating soft female may easily dry out or suffer mechanical injury. Intertidal crabs, such as fiddler crabs (genus *Uca*) and some grapsids, mate when the female is hard. So do certain land crabs. When the female crab is hard at the time of mating, it is possible for her to maintain a position above the male rather than under him, since she does not need the protection of the male above her. In fiddler crabs, for instance, the hard female may remain

Fig. 78. Mating pair of purple land crabs, with hard-shelled female above hard-shelled male. (Courtesy of Jacques van Montfrans and American Museum of Natural History)

only partially beneath the male and, with her last three pairs of thoracic legs, may support the weight of them both.

Scientists have long wondered about how a male crab can tell when a female crab of its own species is ready for mating. Some have speculated that a female crab ready to mate gives off a chemical substance to which a mature male of the same species is sensitive.

The first experimental evidence that such speculation might be well founded came from experiments on the blood-spotted swimming crab, *Portunus sanguinolentus*, by Dr. Edward P. Ryan. When he exposed a mature male of this species to water from a tank containing a mature female about to shed her shell, the male crab began his courtship display, walking about on the tips of his legs, with his body elevated and his claws extended laterally, as he searched for the female. Further experiments by Dr. Ryan indicated that the sex-attractant was released by the female in her urine.

Following Dr. Ryan's initial experiments, numerous investigators have tested other species of crustaceans for the existence of a sex-attractant, or sex pheromone. The most convincing evidence has come from aquatic crustaceans in which the female mates immediately or very soon after shedding her shell. This group includes caridean shrimps and portunid crabs, such as *Portunus sanguinolentus* and the blue crab. Visual stimuli, as well as a chemical sex-attractant, appear to be important in releasing mating behavior in the blue crab. In laboratory trials, Anton R. Teytaud found that immature female blue crabs about to shed their shell at the puberty molt responded by exhibiting their own courtship display when exposed to water from a tank in which a sexually mature male blue crab had been housed, provided that simultaneously a model of a male blue crab, particularly a displaying male, was presented to the female. Mr. Teytaud noted that the crab fishermen, or watermen, of Chesapeake Bay sometimes bait their pots with live male blue crabs when trying to trap female blue crabs about to shed their shell (peeler crabs).

Yet the release of a chemical sex-attractant, or sex pheromone, into the water can obviously be effective only among aquatic crustaceans that mate, or at least seek their

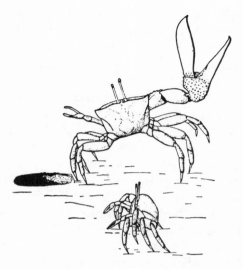

Fig. 79. Courting male fiddler crab of the species *Uca maracoani*, which lives on mud flats on island of Trinidad. Male performs lateral waving of large claw as female approaches, then descends into burrow. Female often responds by following male into burrow and mating there. (By permission of New York Zoological Society)

mate, while submerged. What do intertidal and terrestrial crustaceans do? Answers to this question originated in studies on the fiddler crab *Uca* by Mrs. Jocelyn Crane Griffin. Fiddler crabs are small, gregarious creatures, found abundantly on mud flats and in salt marshes, where at low tide they emerge from their burrows to seek food. The males are recognizable because one claw is greatly enlarged and is often called the "fiddle." Mrs. Griffin found that a male fiddler crab excites a female and prepares her for mating by moving his large claw in certain ritualistic patterns. In some species, mainly of the Indo-West Pacific, the male performs vertical waving motions by raising and lowering his large claw, which remains flexed, in front of his body. In other species, primarily Central and South American, the male carries out lateral waves, unflexing his large claw horizontally outward and then returning it to its flexed position on the same horizontal plane, or circling it up, around, and down again to its original position. Males of still other species may perform waving movements intermediate between these two types. A female fiddler crab may respond to vertical waving by approaching the male and mating with him on the surface of the ground, often at the mouth of his burrow. She usually responds to lateral waving by following the male into his burrow and mating with him there. During courtship, and also during attack and defense, males and females of other semi-terrestrial and terrestrial crabs may perform ritualistic movements of the claws and thoracic legs.

As long as courtship involves visual signals, it can occur only during the day. But male fiddler crabs can produce sounds that enable them to continue courting a female even at night. Most research on sound production has been done on two species, *Uca pugilator* by Dr. Michael Salmon in the United States, and *Uca tangeri* by Drs. Rudolf Altevogt and H. O. von Hagen in West Germany.

During the night, at ebb tide, a male *Uca pugilator* may produce sound almost continuously by rapping his large claw against the sand. During the day, when the male is waving his large claw to attract a female, he may increase his rate of waving and then alternate waving with sound-making or shift to sound-making entirely should a female approach him. If the female then follows him into his burrow, he may make sounds at a very high rate. A male *Uca tangeri* that is waving at a female may shift to sound if the female is temporarily obscured.

In the purple land crab, *Gecarcinus lateralis*, a hard male courts a hard female by tapping with his claws and other thoracic legs on the substrate. In this way, he produces a series of pulses that "tells" the female that he is ready for mating. He also touches her legs and carapace with his thoracic legs. If her ovaries are ripe and thus she is ready for mating, she responds to his courtship by raising her body and allowing the male crab to swing underneath in a position suitable for mating.

Thus, in some semi-terrestrial and terrestrial crabs certain behavioral and acoustical patterns can substitute for the release of a chemical sex attractant to bring a male and a female together for mating. Both the display and the sound-making are "species-specific"; that is, they attract only members of the opposite sex that belong to the same species as does the crab producing the display or sound. Since the chemical nature of crustacean sex-attractants has not been determined, no one yet knows whether these substances are also species-specific.

Six

Development and Growth

Primitively, every crustacean is a creature of the sea, and its early development reflects this ancestral history, even if it lives its adult life in fresh water or on land. As a young crustacean grows and develops, it must change in form. If, as an adult, it will swim freely in the ocean, it has, as it were, only to grow larger and to modify its mouth parts in keeping with a changing diet. But if, as an adult, it will walk on the ocean bottom or inhabit fresh water or roam freely over the land, the young crustacean must transform, sometimes drastically, in order to prepare for a very different adult way of life. A metamorphosis must occur.

At the start, we consider development within the egg. Next, we look at the oceanic larval life of marine crustaceans, including some that spend their entire life swimming freely within the ocean. Then we examine modifications that prepare shrimps, lobsters and crabs for a bottom-living, but still marine, adult way of life. Finally, we see how development and growth are affected in some decapod crustaceans that live not in oceans, bays and sounds, but in fresh water or on land.

Development Within the Egg

The egg of a decapod crustacean starts to develop soon after the penetration of a male sex cell, or sperm. In the shrimp *Penaeus japonicus,* about 30 minutes elapse between entrance of the sperm and the first external evidence of egg division, known as cleavage. In the large king crab, *Paralithodes camtschatica,* this interval is considerably longer, approximating four days. Yet during the intervening period of time, much happens within the egg. Of great importance is the initial event. The nucleus of the sperm unites with the nucleus of the egg to form a fusion nucleus and thereby effect the fertilization of the egg. Each nucleus, through the genes on its chromosomes, contains genetic determinants derived from the parent. Thus, through fertilization, the new organism is provided with its total complement of heritable characteristics.

Eggs of most decapod crustaceans are filled with yolk. As a result, cleavage is superficial; that is, only the outer portion of the egg divides. Total cleavage, in which the whole egg divides, occurs in eggs that have little yolk, equally distributed throughout. Among decapod crustaceans, only the primitive penaeid shrimps undergo total cleavage. Intermediate conditions between total and superficial cleavage exist in certain other groups of decapods. For example, in the king crab and in some caridean shrimps, cell membranes may extend to the center of the egg during the four- and eight-celled stages, but not during later stages of cleavage.

At the start of superficial cleavage, the nucleus of the egg lies in the center of the yolk and is surrounded by a small island of cytoplasm. Male and female nuclei unite, and the resulting fusion nucleus divides into two, the two into four, and the four into eight daughter nuclei. No division of cytoplasm or of yolk usually accompanies this division of nuclei. As the nuclei continue to divide, they migrate slowly toward the surface of the egg. Cell membranes form between the nuclei, but do not reach the center of the yolky interior, thus giving rise to a small, undivided, yolk-filled central blastocoel (after the Greek for "germ hollow," germ being used not in

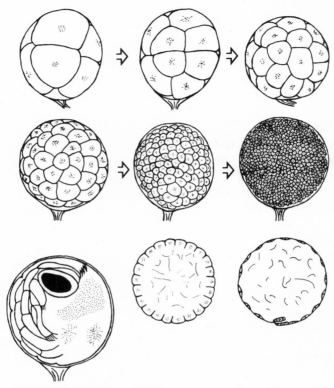

Fig. 80. Developing egg of king crab, egg about 1/32 inch in diameter. From left, top row: 8-cell stage, 16-cell stage, 32-cell stage;
middle row: 64-cell stage, morula (after Latin, *morum*: mulberry),
stereoblastula. Center and right, bottom row: section through
morula and stereoblastula, showing yolk-filled egg, with cells at
surface. Left, bottom row: inside egg, ready to hatch, a zoea (larva),
showing (lateral view) eye (black oval), abdomen (over eye), antennule and antenna (below eye), three maxillipeds (left of antenna), chromatophores (lower stippling), yolk remaining within
egg (upper stippling). Stalk attaches egg to rest of egg mass and to
pleopods of female. (After Marukawa)

the sense of an infectious agent, but with reference to the
developmental potential of the egg). In total cleavage the
blastocoel is large and fluid-filled, and the little yolk that is
present is contained within the surrounding cells.

As superficial cleavage continues, the organism becomes a

single thin layer of innumerable flattened nucleated cells surrounding a solid spherical mass of yolk. The organism is now known as a stereoblastula, which in translation of the Greek words from which the term is derived means "solid germ."

Next the developing organism becomes a gastrula, a term derived from the Greek word *gaster,* meaning "stomach." In primitive decapod crustaceans, such as penaeid shrimps, where the blastula has a large blastocoel containing no yolk, a small oval portion of the external layer of cells becomes thickened and a depression appears within the oval. The depression enlarges and develops into a small-mouthed pouch, the body of which projects into the blastocoel. This pouch is the forerunner of the gut, hence the name gastrula for the organism at this stage.

Before gastrulation, however, most decapod crustaceans are stereoblastulae, comprising a thin layer of cells around a solid mass of yolk. Gastrulation by invagination, as just described, is not possible. Instead, gastrulation takes place by immigration, or ingression. At a point on the blastula, surface cells migrate to the interior. Here they give rise to other cells and eventually to various tissues. For, though the gastrula does not bear any resemblance to the creature that it will become, it has the foundation of all future organ systems.

The subsequent history of the developing decapod crustacean is variable, according to species. In some forms hatching occurs early in development, while in other forms it takes place much later. The longer development within the egg continues, the further advanced in development is the creature that emerges. In certain species, notably those that live in fresh water, the entire early development is spent within the egg and the offspring emerge as miniature—but often somewhat distorted—versions of their parents. In fresh water, delayed emergence has its advantages, as we shall see later in this chapter.

Owing to a series of careful studies by Dr. Charles C. Davis, we now know something of the mechanisms involved in the hatching of a decapod crustacean. Two membranes surround each egg, an outer, relatively thick membrane of soft chitin known as a chorion, and an inner, thinner membrane. In the decapod crustaceans studied by Dr. Davis, which include several species of freshwater and marine shrimps, as well as the American lobster and the blue crab, the hatching process involves, first, rupture of the chorion, second, rupture of the inner membrane, and third, emergence of the young animal.

The chorion ruptures when water enters the inner membrane, as in shrimps and blue crabs, or the enclosed animal itself, as in the American lobster. The inner membrane of a shrimp or blue crab ruptures when the animal struggles to free itself. As for an American lobster, the inner membrane ruptures and the animal emerges as the mother ventilates the eggs she is carrying. She literally washes out her young.

Oceanic Larval Life

When a marine crustacean hatches from its egg, it often looks fantastically different from the adult it will become. Appropriately enough, in this immature state the crustacean is called a larva, after a Latin word meaning "mask." The earliest larval form is known as a nauplius (from Greek *nauplios* or *nautilos,* sailor, or from *naus,* ship), in reference to the free-swimming, generally oceanic habits of the nauplius. Minute and egg- or pear-shaped, the nauplius is unsegmented, with a median eye of two or more parts. The nauplius has only three pairs of appendages; namely, the first and

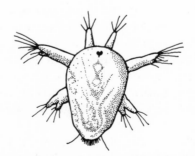

Fig. 81. A nauplius (dorsal view), earliest larval stage of crustaceans; free-swimming in copepods (shown here), penaeid shrimps, and others, but passed within egg in higher shrimps, lobsters, and crabs. (After Green)

second pairs of antennae and a pair of mandibles, or jaws, which at this stage do not appear jaw-like. All three pairs of appendages are used for swimming.

Not all kinds of crustaceans hatch as a nauplius. Among the decapods, the penaeid shrimps do. So also do the branchiopods, ostracods, copepods, barnacles, and euphausiids. Most other crustaceans, including higher shrimps, lobsters, and crabs, pass through the naupliar stage while still within the egg and hatch at a later stage of development.

Upon hatching, a crustacean larva joins throngs of floating, drifting microorganisms that are called collectively plankton. The Greek word from which this term is derived is often translated as "wanderer." However, according to Sir Alistair Hardy, a leading authority on plankton and author of the book, *The Open Sea,* the Greek word is more precisely translated "that which is made to wander or drift." Organisms that constitute the plankton drift or "wander" because they are powerless to do otherwise. Although within the floating mass there are innumerably tiny animals amongst innumerable single-celled green plants, the swimming of these animals cannot offset the steady, relentless drift of the entire mass. Thus their course is determined by the movements of the ocean waters; that is, by the great oceanic currents.

When a crustacean joins other members of the plankton, it joins representatives of virtually every major phylum of the animal kingdom. Most of these phyla are present only as larvae, but adult jellyfishes and marine worms are present. Vertebrates occur as the floating eggs and immature stages of fishes.

Plankton may extend to great depths and as a mass may shift upward during the hours of darkness and downward during the daylight hours, as many species carry out extensive vertical migrations. These migrations appear to be at least partly related to changes in light intensity. Oceanographers have estimated that euphausiids and ostracods migrate vertically more than 300 feet and some other pelagic crustaceans as much as 2,600 feet.

A newly hatched crustacean larva remains in the middle to upper levels of the sea. Here, where sunlight is abundant, so also are single-celled green plants—and so, consequently, is the food supply. However, crustaceans, including the penaeid shrimps, that hatch as a nauplius may not have a functional mouth, or an anus for the egestion of solid wastes. One or more molts may pass before feeding is possible in such forms. In the meantime, the tiny creatures subsist on yolk remaining from the egg. Crustaceans such as higher shrimps, lobsters, and crabs, which hatch in more advanced stages of development, start to feed on unicellular plants and animals of the plankton soon after hatching.

One type of plant predominates in the plankton. It is the diatom, a single-celled green plant with a cell wall of hard, siliceous material that in effect encloses the living portion of the plant in a transparent glass box. Copepods and euphausiids "graze" on diatoms, hence the name often applied to plankton, "pastures of the sea."

The predominant animals of the plankton are copepods, which are present in incalculable numbers and constitute the next step, after diatoms, in food chains of the sea. Hardly larger than the head of a pin, copepods are preyed upon by crabs, worms, mollusks, and small fishes, all of which in turn are attacked and devoured by large fishes, and so on. A few sizable fishes by-pass the longer food chains, with menhaden feeding directly on diatoms, and with mackerel, herring, and shad eating copepods and other small crustaceans of the plankton.

Euphausiids, which are called krill by whalers, are also extraordinarily abundant in the plankton. Each day whalebone whales, such as the blue, fin, and humpback whales, may strain from the water hundreds or thousands of pounds of euphausiids, along with copepods and other planktonic animals.

Euphausiids are particularly good examples of marine crustaceans that hatch from the egg as a nauplius and then gradually develop into an inch-long, free-swimming adult by adding more and more posterior segments and by developing more and more specialized appendages. Among other crustaceans that gradually assume the adult form are some branchiopods, a few species of which live in the sea, and ostracods, which are widely distributed in the sea, as well as in fresh water. At hatching, the nauplius of an ostracod is already enclosed within a bivalve shell similar to that surrounding the adult.

Preparing for Life on the Ocean Bottom

Most shrimps, lobsters and crabs walk on the bottom of the sea when they are adult, but as larvae they are typically free-swimming members of the oceanic plankton. When they change from a larva into a more advanced form called a postlarva, their thoracic appendages become adapted for walking rather than swimming, and on their abdomen appear appendages, known as pleopods, that may be used for swimming. The changes in appearance, as one larval stage progresses into the next, and as the last larval stage transforms into a postlarva and the postlarva into an adult, are not especially impressive in penaeid shrimps and in true lobsters. But these changes are dramatic in spiny, or rock, lobsters and in crabs.

Shrimps

In the white shrimp, *Penaeus (Litopenaeus) setiferus,* larval development is complicated by the fact that eggs need water of high salinity in which to hatch, but postlarvae require water of very low salinity. Hence, as one developmental stage gives way to another, there occurs a migration of sorts, a migration that is dependent upon favorable onshore currents for its completion.

The female white shrimp spawns either within the open waters of the Atlantic Ocean or in the Gulf of Mexico, sometimes fairly close to inlets and sometimes many miles from shore, but always where the water is very saline, with a salt content of about 35 parts per thousand, or 3.5%. This requirement, among others, limits the white shrimp to southern waters or to northward-flowing currents of warm water. The fertilized eggs, which are less than a sixty-fourth of an

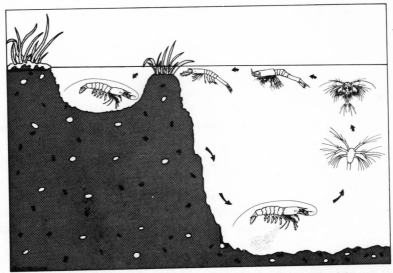

Fig. 82. Life cycle of white shrimp, showing female shedding eggs on bottom of ocean and (counterclockwise) nauplius (ventral view), about 1/100 inch long; protozoea (ventral view), about 1/25 inch long; mysis, about 1/8 inch long; and postlarva, about 5/16 inch long; also young adult, about 5 inches long, in the shallows. (Stages after Pearson)

inch in diameter, disperse and gradually sink to the bottom, where they remain during development.

About 24 hours after spawning, the eggs of the white shrimp hatch. Each larva that emerges is a nauplius about one-hundredth of an inch long. During the following 24–36 hours, with food being supplied entirely by yolk left from the egg, the larva passes through four more naupliar stages, each stage slightly larger and somewhat more complex structurally than the previous one. Owing to the yolk that is present, the nauplii are more or less opaque.

Next the larva changes into a relatively transparent protozoea (pronounced prō-tō-zō-Ē-ă) which is approximately four-hundredths of an inch long. The name of this stage is derived from the Greek, *proto*, first, and *zoōn*, animal, or *zōe*, life. Now the body of the larva is divided into an anterior portion that is covered by a carapace, and posterior abdomen, which becomes segmented in the second protozoeal stage. Also by the second protozoeal stage a forward-projecting beaklike rostrum and a pair of stalked compound eyes appear at the anterior end, while the median nauplius eye disappears. The mandibles lose their bristly branch and become adapted for chewing, but the two pairs of antennae remain bristly and are used for swimming. A digestive tract develops, terminated by a functional mouth and an anus, and the larva begins to feed on the minute floating plants and animals of the plankton.

Finally, the larva, now about one-eighth of an inch in length, closely resembles the adult of a free-swimming, shrimplike but nondecapod crustacean known as *Mysis*, after which this stage is named. By this time the two pairs of antennae have lost their filamentous bristles and have become sense organs of taste and touch. Five pairs of bristly thoracic appendages have appeared, taking over the swimming function. Pleopods, or swimming appendages, have begun to develop on the segmented abdomen.

These various larval transformations require about two or three weeks. During this entire period the larva, swimming weakly, drifts gradually with the plankton toward regions of lower salinity along the shore. As it approaches an estuary, the little shrimp enters the first of several postlarval stages, in which some of its thoracic appendages are transformed into legs for walking and its pleopods begin to be used for swimming. By an advanced postlarval stage the shrimp, about five-sixteenths of an inch long, enters an estuary.

The estuaries, marshes, bayous and lagoons that postlarvae of the white shrimp enter constitute their so-called nursery grounds. Here the water is brackish or almost fresh, sometimes as low in salt content as two or three parts per thousand. Larvae and early postlarvae that are not carried to these grounds by favorable onshore currents perish. On the mud bottom in these areas are worms, small crustaceans, mollusks, and plant debris, which the more advanced postlarvae consume, for by this time the shrimps are bottom dwellers.

Four to eight weeks after entering their nursery grounds, the shrimps, now reaching young adulthood, leave for deeper creeks, rivers, and bays near the ocean. Here they remain through the summer months, feeding on much the same diet as the postlarvae and often reaching a length of five inches before the colder weather of autumn slows their growth. Now the shrimps start toward the warmer, deeper waters of the Atlantic Ocean and the Gulf of Mexico, where ripening of the eggs and spawning take place the following spring.

The development of the young white shrimp from egg to sexual maturity takes one full year, and soon after mating and spawning, the adults usually die. The life span of the white shrimp, therefore, is short, approximating only a year. In exceptional cases, however, a female white shrimp may live for a second year after spawning.

Two other species of penaeid shrimps, the brown shrimp, *Penaeus (Melicertus) aztecus aztecus*, and the pink shrimp, *Penaeus (Melicertus) duorarum duorarum*, both of which also occur off the South Atlantic states and in the Gulf of Mexico, pass through larval and postlarval stages comparable to those of the white shrimp. The same is true of caridean shrimps, except that their naupliar stage is passed within the egg and the larva hatches as a protozoea. In brown and pink shrimps and in some caridean shrimps spawning occurs in oceanic waters of high salinity just as it does in the white shrimp, larvae must be carried toward shore by favorable onshore currents, postlarvae migrate into the brackish waters of bays and lagoons that constitute their nursery grounds, and young adults gradually return to the ocean.

True Lobsters

The American lobster, *Homarus americanus*, like other true lobsters, hatches from the egg at a relatively late stage of development. Hence its larval transformations are less extensive than are those of penaeid shrimps. Indeed, there are only three larval stages in all, and these stages do not differ much in appearance from one another. Only when the young lobster assumes a postlarval condition, preparatory to initiating life on the sea bottom, does a metamorphosis occur. This is also true for the European lobster, *H. gammarus*, and for various species of *Nephrops*, including the Norway lobster, *N. norvegicus*. The larval stages of these lobsters are much like those of the American lobster. Since the postlarvae of true lobsters do not require a salinity considerably lower than that needed by the eggs for hatching, both embryonic and larval development can be completed within the open ocean.

Eggs of the American lobster hatch about a year after spawning. During the intervening period the female lobster, carrying the eggs attached to her pleopods, periodically aerates and cleans them by extending her tail and moving her pleopods back and forth. When the time for hatching arrives, the female stands on the tips of her walking legs with her claws outstretched, her tail held upward at an angle. She waves her pleopods violently, thereby creating a current of water that helps to release the young from their egg cases and to carry them up to the surface of the sea. This procedure is repeated at intervals over a period of about two weeks, by the end of which the entire batch of eggs has been hatched.

On emergence from the egg, which is about one-sixteenth of an inch in diameter, the American lobster corresponds in form to the mysis stage of a penaeid shrimp and is known as a first-stage lobster. It is slightly over one-third of an inch long and a striking creature, transparent as glass and with enormous stalked compound eyes, a conspicuous spine projecting forward from its head, and branched, bristly appendages on its thorax. Because of its transparency, all of its internal organs can clearly be seen. The larva uses the outer branch of each thoracic appendage as an oar to move its segmented body upward and forward, and its segmented abdomen to dart quickly backward.

The American lobster passes through two more mysis-type larval stages, known, respectively, as the second-stage lobster and the third-stage lobster. In the second-stage, all parts of the body are larger, with the exception of the swimming appendages on the thorax, and several pairs of pleopods have appeared on the abdomen. A third-stage lobster is even bigger, having now reached about one-half inch in length, and it has gained a pair of relatively large claws. Although the

Fig. 83. Life cycle of American lobster, showing spawning female on ocean bottom (lower right) and (clockwise) female releasing larvae from egg cases on pleopods; first-stage lobster (first larva) and second-stage lobster (second larva), both about 1/3 inch long and both shown swimming forward and upward; third-stage lobster (third larva), about 1/2 inch long and shown swimming backward and upward; fourth-stage lobster (postlarva), about one inch long. (Immature stages after Herrick; adult stages after photos by Wilder)

thoracic appendages are still used for swimming, the abdominal appendages are present almost in full array and are fringed with bristles (setae), ready to be used as swimming organs. From somewhat less than one week to somewhat more than two weeks are required for a larva to pass through these three developmental stages. During this period of time it drifts with the plankton, consuming creatures tinier than itself.

On entering its fourth developmental stage the American lobster undergoes a marked transformation. The swimming branch on each thoracic appendage becomes reduced to a nonfunctioning stump, while the remaining branch turns into a full-fledged leg. The shell loses much of its transparency and becomes multicolored. Indeed, the young lobster, now a postlarva, has acquired a striking resemblance to an adult. But it is only an inch or so in length, and it still swims at the surface of the sea, using its pleopods to move rapidly forward and its abdomen to dart backward. In defending itself, the lobster brandishes its large claws, much as does an adult. Toward the end of the postlarval stage the lobster swims less and less at the surface and finally sinks to the sea bottom, where henceforth it will spend its days. In locomotion and general habits, the young lobster now strongly resembles an adult, even to showing preference for a rocky bottom.

A full year and many more molts elapse before an American lobster reaches a length of one or one and a half inches, and three years before it is three to four inches long. As it continues to grow larger, it feeds on bottom-living animals, such as fishes, starfishes, worms, clams, mussels, sea urchins, and crabs. The size of a lobster at the time of sexual maturity depends upon the temperature of the water in which the lobster is living, since a higher temperature hastens growth and sexual maturity. For example, in relatively warm waters like those of the southern Gulf of St. Lawrence, sexual maturity may be reached when a female is no more than seven inches long and weighs less than a pound, but in colder waters such as those of the Bay of Fundy, a female lobster may not reach maturity until she is 12 inches long and weighs about two pounds.

The life span of an American lobster may be short, mainly because man is such a persistent and efficient enemy. But if a lobster can escape the clutches of man and other predators, like the catfish, dogfish, codfish, and skate, it may live to a ripe old age. With the advent of a fishery for lobsters from George's Bank southward, 10–15-pound giants are being caught regularly. These lobsters may be 15 or 20 years old. Some are larger and much older.

Spiny Lobsters

The characteristic free-swimming larval stage of the spiny,

or rock, lobsters and of the slipper lobsters is a flattened, transparent, leaf-like form known as a phyllosoma, the name being derived from the Greek words *phyllon*, leaf, and *soma*, body. The larva is suitably named, for its anterior portion consists of two leaf-like regions, one its so-called cephalic (head) shield and the other its thorax. At the time of metamorphosis into a postlarva, the two anterior leaf-like regions fuse into one more or less rectangular region, the cephalothorax. Due to its leaf-like shape, a phyllosoma is admirably fitted for a long planktonic existence, some species being able to sustain pelagic, or open sea, life, with periodic molting, for up to six or seven months. During this time the phyllosoma may travel great distances.

Corresponding in some ways to a protozoea and in other ways to the mysis stage of a penaeid shrimp, the phyllosoma formerly was thought to be the form into which the eggs of spiny lobsters hatch. But two earlier free-swimming stages, the prenaupliosoma and the naupliosoma, have been described for the South African rock lobster, *Jasus lalandii*. Both prenaupliosoma and naupliosoma are more complex in structure than the nauplius, a stage that a spiny lobster always passes within the egg.

The fertilized eggs of the South African rock lobster, which are approximately one-sixteenth of an inch in diameter, hatch about three months after spawning. During the interim, the eggs undergo embryonic development within the egg cases, carried by the female on her pleopods. At hatching, there emerges a prenaupliosoma that within eight hours molts into a naupliosoma, less than one-sixteenth of an inch long. The naupliosoma has two large stalked compound eyes and a pair of long, well-developed antennae that bear filamentous bristles, or setae. These feathery structures are used for swimming, while three pairs of long thoracic legs are kept folded against the body.

After approximately four to six hours, the naupliosoma molts into a leaf-like phyllosoma, which is slightly over one-sixteenth of an inch long. Now the bristles are absent from the antennae, while a bristly outer branch of the first and second thoracic legs takes over the swimming function. The body,

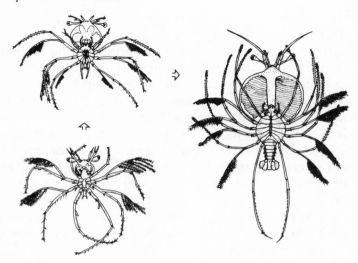

Fig. 84. Transparent, pelagic phyllosoma (larva) of South African rock lobster, shown (clockwise from lower left) after molting to second, fifth, and thirteenth (final) stage. Increase in size, from about 1/16 inch to just over 1 inch. (After Lazarus)

which was fairly opaque during the preceding stage because of the yolk that was present, is transparent. With all its yolk gone, the phyllosoma feeds on plankton.

Toward the end of the phyllosoma stage, the larva may leave the surface layers of the ocean and go to the bottom, although there is some doubt in the minds of scientists as to just when in the life cycle this change in habitat takes place. In any event, through successive molts the phyllosoma continues to grow larger and larger, eventually reaching more than one inch in length.

A truly striking transformation takes place as the phyllosoma changes into the postlarval stage, called a puerulus (after a Latin word for "child"). As a puerulus, the rock lobster is still transparent but otherwise looks very much like an adult. The most noticeable change has been in the shape of the body, which now consists of a roughly rectangular cephalothorax and a long abdomen. The thoracic appendages have lost their hairy branch and are strictly for walking on the bottom, while pleopods have developed as feathery, two-lobed appendages useful for swimming. The rock lobster now is slightly less than one inch in length, having decreased somewhat in size during metamorphosis from a phyllosoma.

Subsequently, deposits of calcium salts appear in the shell of the young rock lobster, hardening it and ridding it of its transparency, but further growth appears to be very slow. A one-inch long rock lobster was observed in the laboratory over a period of two years, during which the animal grew only about a quarter of an inch. Possibly confinement slowed its rate of growth, which in nature might have been somewhat higher.

As a young rock lobster continues to grow and molt and approach adulthood, it lives and feeds on a rocky bottom. Its only means of swimming now is its tail, which by sudden flexion can cause the lobster to move suddenly and jerkily backward through the water. It has very strong mandibles, or jaws, and these apparently are used as "nutcrackers" to crack open the shells of the various kinds of mollusks that serve as a principal item of food. The rock lobster may also feed on seaweed that generally abounds in its habitat and provides cover, although some rock lobsters may be found on a bare, and occasionally even on a sandy, bottom. The life span of a South African rock lobster is unknown, but individuals that have been reared in captivity from the larval stage have lived as long as ten years.

This description of growth and development in the South African rock lobster is based largely upon observations reported in 1935 and 1936 by Dr. Cecil von Bonde. In 1956 Dr. Martin W. Johnson described larval and postlarval stages of the California spiny lobster, *Panulirus interruptus,* and in 1962 Dr. R. W. George described those of the Western Australian rock lobster, *Panulirus longipes cygnus.* Only in their details do the larval stages of these species differ from one another.

Anomuran Crabs

In both anomuran and brachyuran (true) crabs, the first free-swimming larval stage is a zoea, corresponding to the

Fig. 85. Adult South African rock lobster, which typically lives on rocky bottom near shore and feeds mainly on mollusks. (Courtesy of the American Museum of Natural History)

mysis stage of a penaeid shrimp. Earlier developmental stages of these crabs are passed within the egg. Typically, a zoea has one or more long spines projecting forward or backward from the carapace and a slender curved abdomen that may terminate in a forked segment, the telson. As in the mysis stage of other crustaceans, the thoracic appendages bear filamentous bristles, or setae, and are used for swimming. In addition, the zoea can progress in a more or less convulsive manner by quick repeated jerks of its abdomen.

From a zoea an anomuran or brachyuran transforms into a postlarva that bears a considerable resemblance to an adult crab, except that the abdomen retains appendages and is stretched out to the rear as in a lobster. Its stalked compound eyes are very conspicuous.

In pagurid anomurans, such as the king crab and hermit crabs, the postlarva is called a glaucothoë. This term is derived from a species of pagurid postlarva, *Glaucothoe peronii*. The Greek word *glaukos*, upon which the generic name is based, means "silvery" or "bluish-gray," presumably a reference to the appearance of the animal. Postlarvae of the species *Glaucothoe peronii* live in the deep sea and have been caught in nets during hauls, but apparently no one has yet collected the young crabs into which the postlarvae are presumed to metamorphose—or at least if the crabs have been collected, no one yet realizes their relationship to the postlarvae of *Glaucothoe peronii*.

The glaucothoë of the hermit crab *Pagurus* is free-swimming at first but, before molting to a second postlarval stage, the glaucothoë takes over a discarded mollusk shell and starts living in it, just as does an adult hermit crab.

The female of the king crab, *Paralithodes camtschatica*, carries fertilized eggs attached to her pleopods throughout their development. Each egg, which is about three-hundredths of an inch in diameter, changes in color from blue-violet to red-orange as it develops. Hatching does not take place until almost one year after spawning. Shortly after hatching, another mating and spawning occur, with the result that a female king crab is without developing eggs for only a few days or a few weeks each spring. Although new eggs start to form in the ovaries as soon as a batch of ripe eggs starts down the oviducts during mating, the new eggs remain within the ovaries for an entire year, ripening only in time for the next mating season in the following spring.

During hatching of the eggs, which may continue for five days before it is complete, myriads of tiny, free-swimming zoeas escape from the egg cases. The zoea has a long forward-projecting beaklike spine and two short, stout spines directed to the rear. The two compound eyes are large, the abdomen is long, curved, and segmented, and the thoracic appendages have long bristles and are used for swimming. At the time of hatching, the zoea is about the size of the head of a pin.

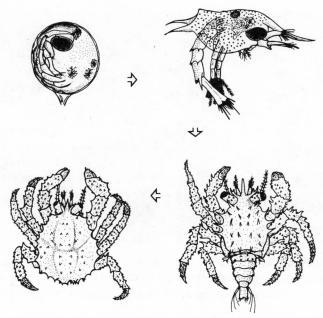

Fig. 86. Early developmental stages of king crab, showing (clockwise from upper left) zoea (larva) within egg case and about to hatch; free-swimming first zoea, about 1/25 inch wide; glaucothoë (postlarva), about 1/16 inch long; first crab stage, slightly less than 1/8 inch wide. (After Marukawa)

After two or three months and three or four more zoeal stages, the zoea metamorphoses into a postlarva, the glaucothoë, just over one-sixteenth of an inch long. At this stage the forward portion of the animal appears crab-like, but the abdomen remains stretched out in back and pleopods are used for swimming. At the following molt, which occurs in about two weeks, there appears a minute crab that is still less than one-eighth of an inch in width. The abdomen is now carried, crab-fashion, tucked underneath the cephalothorax.

During its weeks as a zoea, the crab lives in the middle and bottom vertical zones of the sea, in water that varies in depth from 120 to 260 feet. On becoming a postlarva and then a tiny crab, however, the animal begins to forage exclusively in the bottom zone, amongst a luxuriant growth of seaweed,

Fig. 87. Female king crab shaking dark egg mass in "brood pouch" under abdomen and thereby helping zoeas (larvae) to escape from egg cases. (© National Geographic Society, Robert Sisson, photographer)

sponges, and a type of encrusting colony of animals known as bryozoans.

Young king crabs molt eight times during their first year and probably four times during their second, then less frequently until they are molting once a year when four or five years old and once every other year when six or seven years old. Their diet consists largely of mollusks, brittle stars, sea urchins, and starfishes, but it also may include various kinds of crustaceans, certain segmented worms known as polychaetes, and seaweed.

The rate of growth of a young king crab is relatively slow. Scientists have estimated that the carapace width may be only three-eighths to one-half inch by the time the crab is one year old, one inch or more when two years old, and one and one-half to two inches when three years old. The crab becomes sexually mature when it is three and one-half to four inches in carapace width, weighs up to two pounds, and is five to six years old. It can be taken legally by fishermen when it is seven inches in carapace width—or about three feet from

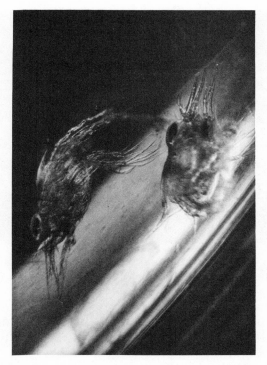

Fig. 88. Swimming zoeas (third-stage larvae) of king crab, approximately two weeks after hatching and still about the size of a pinhead. (© National Geographic Society, Robert Sisson, photographer)

claw to claw. The crab now weighs six to seven pounds and may be seven to ten years old. If a king crab escapes capture, it may live for 20 to 25 years, possibly attaining a weight of 25 pounds or more and a spread of over six feet.

True Crabs

Although the larva of a true, or brachyuran, crab shares with the larva of an anomuran crab the name of zoea, the postlarva of a brachyuran crab has its own distinctive name, megalops—or megalopa. The Greek word *megas* upon which the name is based means "large" and may refer to the very

Fig. 89. Juvenile king crabs, shown at about one-fourth natural size. (Courtesy of the American Museum of Natural History)

conspicuous stalked compound eyes that a megalopa posses-ses. As is the case for an anomuran postlarva, a brachyuran postlarva is crab-like in appearance, except for its abdomen, which stretches out to the rear in lobster-like fashion and retains several pairs of pleopods.

Without their realizing it, some people have had uncom-fortably close association with the megalopa of the blue crab, *Callinectes sapidus*. On occasion this postlarva has appeared in large numbers near the shore and in the breakers at Virginia Beach, Virginia, close to the mouth of Chesapeake Bay, where the postlarvae have bitten swimmers and bathers and caused complaints of "water fleas." The megalopa of the blue crab is approximately flea-sized, being slightly over one-eighth of an inch long.

Both the megalopa and the zoea of the blue crab require water of high salinity. Yet the adult of this crab mates in water of very low salinity. Hence the development and growth of the blue crab, like that of penaeid shrimps, is accompanied by a migration. In the case of the blue crab, the migration extends from far up in estuaries like Chesapeake Bay down into the waters of the Atlantic Ocean and back again up into the estuaries. Although the following account specifically comcerns blue crabs in Chesapeake Bay, in essence it describes the development of blue crabs in many other estuaries along the Atlantic and Gulf coasts.

Mating in blue crabs takes place immediately after the female has shed her shell for the last time and while she is still soft. During this shedding, which is known as her terminal ecdysis, her abdomen loses the triangular shape that it has had up to this time and becomes broadened and rounded, capable of brooding the eggs that she will produce. The female now is known to fishermen as a "sook," which is a distortion of the word "sow." Before maturity, female blue crabs are often called "Sally" crabs.

Once mating is complete and the female has hardened, she begins her journey down Chesapeake Bay, feeding on fish, shellfish and seaweed as she goes. If mating has occurred early in the summer, the female probably reaches the lower Bay by autumn. Other females, mating later in the summer, may spend the winter in the middle Bay and resume their journey in the following spring. The female crabs that do complete their southward migration by late autumn either congregate in the mud near the mouth of Chesapeake Bay, where they constitute the winter dredge fishery, or pass between Cape Charles and Cape Henry out into the Atlantic Ocean. Male crabs, which often are referred to as "Jimmies," remain in the upper Bay.

Spawning begins early in May and continues through much of the summer. Hatching of the two million or so eggs that constitute the "sponge" of a female blue crab takes place about two weeks after spawning. Out of each viable egg, which is about one-hundredth of an inch in diameter, comes a prezoea. Within the next one to three minutes the first larval ecdysis occurs and a full-fledged zoea struggles free to swim vigorously away.

A prezoea looks rather like a shorn zoea because it appears to lack the conspicuously long rostral and dorsal spines, the short lateral spines, and the long, feathery thoracic appendages that are characteristic of a zoea. Actually the spines and

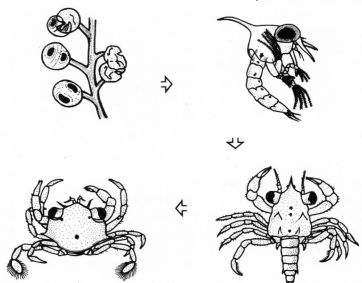

Fig. 90. Early developmental stages of blue crab showing (clockwise from upper left) eggs, approximately 1/100 inch in diameter and about to hatch or hatching (empty egg cases at right); zoea (larva), about 1/25 inch long; megalops (postlarva), about 1/8 inch long; and first crab stage, about 1/10 inch wide. (After drawings by Roy L. Robertson)

appendages are present but they are kept bent against the body of the zoea by the prezoeal covering, or cuticle, that ensheaths the zoea. As soon as this cuticle is ruptured, the spines and appendages of the zoea stretch out into their normal position.

The zoeas, feeding abundantly on microscopic plants and animals of the plankton, undergo four or more ecdyses and grow to be about one twenty-fifth of an inch in length before transforming into a megalopa. Development up to this point has taken from five to seven weeks. These tiny blue crabs remain in the very saline waters of lower Chesapeake Bay and nearby coastal waters. Although they may be found swimming at the surface, they sink to the bottom as soon as swimming ceases. Life as a megalopa lasts for only a few days, after which the blue crab enters the so-called first crab stage. It is now about one-tenth of an inch in width and looks like a crab.

By the early part of August a sizable number of blue crabs have reached the first crab stage, and a northward migration of the new generation begins. According to W. A. Van Engel of the Virginia Institute of Marine Science, the first migratory wave of young blue crabs reaches the rivers along the Virginia shoreline of Chesapeake Bay around the third week of August. By September and October, crabs that are one-quarter to one-half of an inch in width are commonly seen in this region. Usually the small crabs can migrate no farther north than the mouth of the Potomac River before cold weather sets in, and thus many spend the winter months in Virginia waters. The northward migration is resumed in the next spring, with crabs from one-half inch to one inch in width appearing in the upper reaches of Chesapeake Bay by late April or May of the second year.

During their migration northward blue crabs molt frequently, the interval between ecdyses increasing as the crab grows older. Crabs about one-fifth of an inch wide may shed every three to five days, those one-half to one inch wide every 10 to 15 days, and those four inches or more in width every 20 to 50 days. Altogether a blue crab molts about 19 times after completing larval and postlarval development. Since molting takes place largely in shallow water, it makes possible the soft-shelled crab industry for which Chesapeake Bay is famous. Blue crabs are about one year old when they reach a width of three inches, which is the minimum legal size for marketing.

A blue crab spends from a year to a year and a half developing from egg to mature adult. Once adulthood is reached, a blue crab may live for at least one more year. Some hardy individuals are known to have lived for three and one-half years.

For a marine brachyuran, the blue crab is unusual in that the adults require—or prefer—water of low salinity, while the early developmental stages, like those of other marine brachyurans, require water of high salinity. A migration, even more extensive than that of the blue crab, is made by the Chinese crab, *Eriocheir sinensis,* at the time of spawning. Since the Chinese crab penetrates far up rivers, however, it usually is classified by scientists as a freshwater rather than a marine

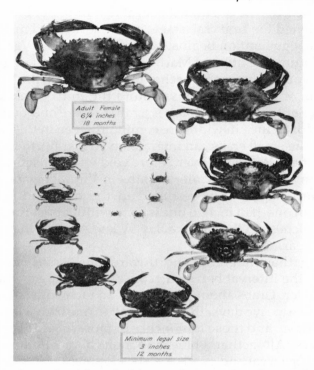

Fig. 91. Growth of female blue crab from tiny first crab stage to mature adult about 1 1/2 years and 19 molts later. Minimum legal size for taking peeler crabs is 3 inches in carapace width, soft crabs 3 1/2 inches, and hard crabs 5 inches. (Courtesy of the Virginia Institute of Marine Science, Gloucester Point, Virginia)

crab. In the remaining brachyurans, a large-scale migration does not accompany development and growth, because the adults inhabit water of salinity fairly close to that required by the eggs, zoeas, and megalopas.

Growing Up in Fresh Water

Most freshwater crustaceans complete their larval development while still within the egg, which is large and has a generous amount of yolk. Since the egg is quite impermeable to salts and water, the larva of a freshwater crustacean thereby avoids the major problem of life in fresh water, that of maintaining a proper balance of salts and water in its tissues. When the egg hatches, the animal that emerges is a miniature version of its parents, with salt- and water-regulatory mechanisms like those of its parents.

The water fleas (Cladocera) are among the freshwater crustaceans that go through their larval stages within the egg, as do freshwater isopods, freshwater amphipods, some freshwater shrimps, freshwater crayfishes and almost all freshwater crabs.

A young freshwater crayfish looks much like its parents when it climbs out of the ruptured membrane of its relatively large egg. Its eyes are huge, its carapace hump-backed, and its claws very slender, but it has most of the appendages of its parents, lacking only the first pair of pleopods and the uropods, which, together with the terminal flap, the telson, make up the tail fan.

After hatching, the young crayfish may remain for a time attached to its parent by the inner membranous covering of the embryo, which adheres as a filament to the young crayfish's telson. When this filament breaks, the crayfish clings to the pleopods of its parent by means of certain claws, which have a small hook at their tip. In the superfamily Astacoidea, made up of crayfishes of the Northern Hemisphere, such modified claws occur on the first pair of thoracic legs, whereas in the superfamily Parastacoidea, the crayfishes of the Southern Hemisphere, a hook occurs on the claw that terminates the fourth and fifth thoracic legs.

After clinging to the female for a week or two and undergoing several molts, the young crayfish leaves the parent, and begins to shift for itself. It continues to grow at a rapid rate, undergoing several more molts during the first summer, then fewer during subsequent growing seasons. A freshwater crayfish generally lives to be three or four years old.

The eggs of freshwater crabs are much larger and fewer than those of a marine crab. Thus, each egg of a freshwater crab may be one-tenth to two-tenths of an inch in diameter, while that of a marine crab is closer to one-hundredth of an inch in diameter. A single female marine crab may produce from 3,000 to 2,000,000 eggs at a single spawning, according to her species, while the number of eggs produced by an individual female freshwater crab at one spawning may vary from 50 to 500, according to the species. Like a marine crab, a female freshwater crab lays her eggs within a "brood pouch" formed by the ventral surface of her thorax, her flexed abdomen, and her pleopods, to which the eggs become attached.

The entire larval development of almost all freshwater crabs takes place within the egg. When the eggs finally hatch, tiny crabs emerge, looking much like their parents despite the fact that they are only one-eighth to one-quarter of an inch in width. The newly hatched crabs cling to their mother for some days before leaving her to begin a life of their own. If the region inhabited by the crabs is low lying and subject to monsoonal rains, the onset of these rains is the signal for the young to leave their parent. They can be seen on the bottom in flooded areas, feeding on small animals and plants. As the flood waters recede, the little crabs burrow into the mud where they remain, emerging from their burrow to seek food.

On the island of Jamaica a species of crab, *Metopaulias depressus*, of the family Grapsidae, lives in rain water that collects in certain flowering plants of the pineapple family known as tank bromeliads. The lower portions of the leaves overlap to form a number of small niches, in which rain water accumulates and which, together with a central reservoir, constitute the "tank."

The eggs of the bromeliad crab are bright orange when newly laid and quite large, measuring about six-hundredths of an inch in diameter. They also are relatively few in number, 60–100 eggs in a single sponge. The first two larval stages, that is, the zoeal stages, occur not within the egg but within water collected by the bromeliad. Yet these stages are abbreviated and are completed within three days. During this time, the larvae consume only yolk remaining from the egg. In the final, postlarval stage, the megalopa, some yolk remains, but the young crab feeds actively. The postlarval stage persists for six to seven days, following which the animal enters the first crab stage.

The bromeliad crab is an example of a freshwater crab that does not undergo its entire development within the egg. The larvae of this crab must face the problem of maintaining a normal salt and water balance within their tissues. So also must the larvae of certain freshwater shrimps, which, unlike many of their kind, have small eggs and free-swimming freshwater larval stages. How these larvae maintain their internal balance of water and salts is not known.

There is a very effective way in which freshwater crustaceans can overcome the problems of larval life in fresh water. That is to avoid them entirely, by migrating to bays and estuaries before the young are released. Females of some freshwater shrimps, including the North American species *Macrobrachium carcinus,* carry their numerous small eggs downstream into brackish water before the eggs hatch into the first of several free-swimming larval stages.

Preparing for Adult Life on Land

The young of crustaceans that, as adults, spend much or almost all of their time on land may go through developmental stages either in the ocean or in fresh water. If the adults return to the ocean to spawn, the young develop in sea water. The eggs produced and the number of developmental stages through which the young pass are similar to those of strictly marine crustaceans. If the adults return to fresh water to spawn, the young develop in fresh water and, as in more aquatic species, generally remain within the relatively large egg while they do so.

A familiar terrestrial crab of the southeastern Atlantic coast is the great land crab, *Cardisoma guanhumi.* Like other members of its family, the Gecarcinidae, it returns to the sea at spawning time. And like strictly marine crabs, the female of this species produces numerous small eggs, which hatch into free-swimming zoeas. After several zoeal stages, the larvae enter a postlarval stage, the megalopa, which crawls out on land and metamorphoses into a tiny crab.

Fig. 92. Large numbers of the great land crab, *Cardisoma guanhumi,* **at Homestead Bayfront Park in southern Florida during spawning migration to the sea, where the females release their larvae from egg mass attached to their abdominal appendages. (Courtesy of Jack Stark)**

This land crab is unusual in that the adults live in large colonies, often several miles from the sea. Although burrows of these crabs extend down to ground water, the water is often quite fresh, and thus unsuitable for the development of the young. Consequently, at spawning time, the females undertake a mass migration to the seashore.

According to Dr. Charles Gifford, the spawning period extends from late June to December. A female carries between 300,000 and 400,000 eggs, each less than two-hundredths of an inch in diameter and, as she carries them, they slowly develop. After about ten days, they have reached the prezoeal stage and are ready to hatch. In the meantime, many hundreds of other spawning females in the colony have been carrying developing eggs, and these eggs also are now ready to hatch.

Just before full moon, the spawning migration takes place and the females march to the sea shore. Here they seek cover near the edge of the water, entering the water frequently and rapidly fanning their pleopods, thereby helping the young to escape from the egg mass. Optimal salinity for hatching is about 14 parts per thousand, or 1.4%, which represents fairly brackish water. In southern Florida the main spawning period occurs at the height of the rainy season, when considerable run-off may greatly reduce the salt concentration of littoral waters.

After hatching, the young of the great land crab go through five zoeal stages, requiring somewhat over three weeks, and enter the megalopa stage, which lasts for about two and a half weeks. Then the young enter the first crab stage.

The larvae of these crabs are typically planktonic in their adaptations and are unable to withstand extremes of salinity and temperature. But the first crab stage and subsequent early crab stages can grow and molt in waters having a wide range of salinities—from zero to an extremely saline 40 parts per thousand. This seems prophetic of things to come, for juveniles and adults of this species seem equally at home in fresh and saline water. This tolerance of the older crabs is an important adaptive feature, enabling the crabs to use ground water as a source, whether fresh or saline.

Other gecarcinids, notably the Caribbean species *Gecarcinus lateralis* and *Gecarcinus ruricola*, also must migrate to the sea to release their young, for although the adults of these species may live some distance from ocean, bays, or estuaries, the young can develop only in sea water. An early twentieth century British zoologist, W. T. Calman, has written of the black land crab, *Gecarcinus ruricola*, as follows: " . . . the migration to the sea takes place annually during the rainy season in May. The Crabs are described as coming down from the hills in vast multitudes, clambering over any obstacles in their way, and even invading houses, in their march to the sea."

The purple land crab, *Gecarcinus lateralis*, does not burrow down to ground water. On the contrary, this land crab digs its burrows in quite dry soil, depending upon dew and often infrequent rainfall to provide moisture. In areas where rain is plentiful, the crab may live in nooks and crannies and under rocks, coming out in great numbers immediately following a shower. The crabs may be seen outside of burrows or nooks and crannies during the day, providing the area offers dim light, warmth, and dampness. Generally, however, the crabs

Fig. 93. Purple land crab carrying load of sandy soil out of burrow, entrance at lower left. (Courtesy of the American Museum of Natural History; Jacques van Montfrans, photographer)

Fig. 94. Purple land crabs (at least a dozen visible) at Sabal Point, Boca Raton, Florida, clinging to strangler fig trees after emerging from burrows, nooks and crannies following heavy rain. (Courtesy of the American Museum of Natural History; Jacques van Montfrans, photographer)

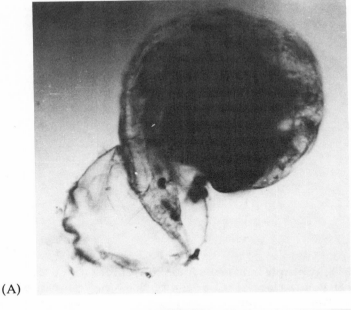

(A)

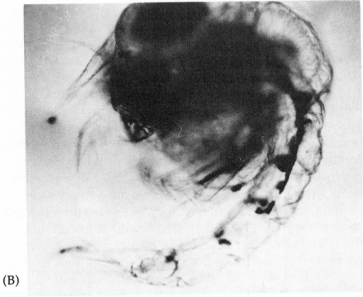

(B)

Fig. 95. Zoea (larva) of purple land crab, photographed in sea water while escaping from ruptured egg case (A) and swimming freely (B). Actual size of zoea, less than 1/16 inch. [Photographs by author; (B) reproduced through courtesy of the *American Zoologist*]

Fig. 96. Tiny purple land crabs found under rock near ocean beach at Sabal Point, Florida. (Courtesy of the American Museum of Natural History; Jacques van Montfrans, photographer)

are nocturnal in their activity, seeking food only after dusk has fallen and returning to shelter by dawn. The purple land crab mates either inside or outside its burrow, carries its eggs while they develop, and migrates to the sea to release its young. After going through zoeal and megalopa stages in the ocean, the young crabs can be found on land, usually hiding under rocks or in detritus not far inland.

Among crabs that return not to the sea but to fresh water to hatch their young is the common land crab *Paratelphusa (Barytelphusa) guerini* of Salsette Island, off the western coast of India, just north of Bombay. According to an account by zoologist Charles McCann in the *Journal of the Bombay Natural History Society* in 1938, the island was infested with this species. The crabs dig burrows in paddy fields or, when the fields become flooded by the rains of the summer monsoon, just above the water level in nearby embankments. As the water level falls, the crabs excavate their burrows farther down on the banks, until at last the crabs are burrowing directly into the stream bed. The location of each burrow is apparent from a large mound of earth, called a "castle," that is formed as the crab carries up pellets of wet mud and heaps them on top of one another near the entrance of the burrow.

While the crabs are making their burrows in the rapidly drying fields and stream beds, they collect leaves, grasses, and other food materials, which they store in the burrow. Once the land surface has become quite dry, these crabs retire into their burrow and plug the entrance from within.

Nothing further is seen of these crabs until the monsoon rains return the following summer. Then out come the crabs from their burrows, the females laden with young—as many as 250 young per female. Apparently mating, spawning, embryonic development, and hatching all take place while the crabs are secluded underground within their burrows, some of which presumably are interconnected.

The young, each somewhat less than one-eighth of an inch in width, are almost fully developed little crabs, essentially miniatures of their parent. They form a compact mass under her flexed abdomen. As rain continues, puddles form, and streams begin to flow, the females with their young congregate on the banks and in the water where they spend much time submerged. The young cling to their parent for a few days, after which she sheds them or they leave of their own accord. Now the countryside teems with little crabs. They spend lots of time in small streams, feeding on algae and minute organisms that live on or near the bottom. In contrast, the adults become less and less conspicuous, although where they go no one seems to know. Possibly the old crabs die—after completing a very short life span of only a year or so.

Seven

Molting

In its simplest terms, molting in crustaceans is the periodic shedding of the hard outer covering, the old exoskeleton, or shell. But this concept is overly simplified, because shedding an old shell requires the prior laying down of a new soft one under the old and, subsequently, the hardening of the new one into a firm, resistant, useful outer covering. The term molting, then, implies a large amount of physiological activity both before and after the actual shedding of the old shell. We shall examine some of this activity and determine what switches it on at the proper moment and off again at a later time.

Although shedding of the shell is not the only event to occur during molting of a decapod crustacean, it is by far the most conspicuous one. Shedding, known technically as ecdysis, occurs periodically, the event taking hardly longer in a large, older animal than in a small, younger one, but the interval between successsive ecdyses gradually increases with age. This is shown in the accompanying figure, which illustrates successive cycles in the ascending and widening spiral of growth. The events that precede ecdysis (proecdysis; *pro*:before), those that follow ecdysis (metecdysis; *met:* after), and

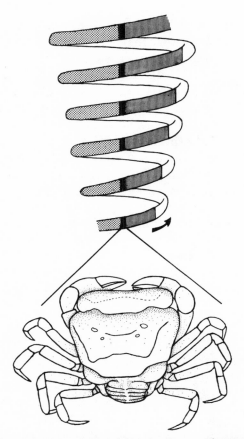

Fig. 97. Successive cycles of molting in decapod crustaceans. Any given cycle includes: proecdysis (dotted area), ecdysis (solid black area), metecdysis (lined area), and anecdysis (clear area). Figure below shows purple land crab withdrawing from old shell during ecdysis. (After Bliss and Boyer)

those that occur between ecdyses (anecdysis; *an*: without) all take increasing amounts of time, but ecdysis itself takes about the same amount of time. The illustration also shows how little time ecdysis requires even in the shortest cycle, compared to the amounts of time needed for the other three portions of the cycle.

Anecdysis has been defined as a long period of rest between the end of one metecdysis and the beginning of the next proecdysis, and it occurs in animals that molt seasonally. But some decapod crustaceans molt all year round, and in these there is no anecdysis. Instead, there occur diecdyses (*di:*two), short periods during which one metecdysis passes almost indiscernibly into the next proecdysis. In some crabs, a series of molts, separated by diecdyses, take place during the summer months, with the whole series followed by anecdysis as the crabs endure the long winter.

Preparations for ecdysis (i.e., proecdysis) begin when a shrimp, lobster or crab starts laying down under its old exoskeleton a new one, complete in every detail, including the pigments that give distinctive color and pattern to the shell. At the same time, the animal starts to regenerate new limbs in place of those that it may have lost since the previous ecdysis. As proecdysis continues, the animal also withdraws certain important mineral salts from its old exoskeleton and stores them in its tissues. If the crustacean lives in fresh water or on

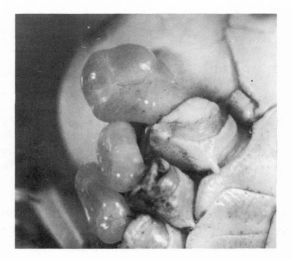

Fig. 98. Regenerating limbs (ventral aspect) of purple land crab. Center top, limb bud of right claw (first thoracic leg); below, limb bud of right second and right third thoracic legs. (Photograph by author; reproduced through courtesy of Zoological Institute, Lund, Sweden)

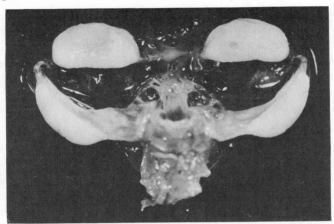

Fig. 99. Excised foregut of purple land crab, showing four gastroliths on framework of anterior chamber and shiny, soft chitinous lining of foregut. (Courtesy of *American Zoologist*; Frank White, photographer)

land, it stores salts much more extensively than if it lives in the sea, where minerals abound. In some species of decapod crustaceans, including freshwater crayfishes, certain land crabs, and the American lobster, stored calcium salts appear as hard, white, oval deposits that form on the hard framework of the anterior chamber of the foregut. These deposits are known as gastroliths (from the Greek, *gaster:* stomach; *lithos:* stone). As ecdysis approaches and the framework of the foregut becomes less rigid, the gastroliths remain loosely held in place by the chitinous lining of the foregut. This lining is shed at ecdysis, and the gastroliths fall into the anterior chamber, where they are soon dissolved by the digestive juices. The freed calcium salts then pass into the hemolymph and subsequently are redeposited in the new soft shell, which they help to harden.

But we have gotten ahead of our story, which is concerned with what is occuring during proecdysis. By the time the decapod crustacean is within a few days of ecdysis, certain areas of its exoskeleton, where reabsorption has occured most extensively, are very soft. One such area is the epimeral (ĕp-ĭ-MĒ-răl) suture, which runs around the crab from front to back and is the line along which the exoskeleton splits when ecdysis occurs. Another is a small area at the base of the claw, which softens enough for the large claw to be withdrawn at ecdysis. When this area does not soften sufficiently, an animal may be unable to withdraw its claw and may reflexly break it off at a preformed breakage plane near the base in a process known as autotomy.

During ecdysis, a marine or freshwater decapod crustacean takes in water, often drinking it and then absorbing it through the walls of the gut, which at this time are very permeable. The animal uses this water to stretch its new soft exoskeleton, often enlarging it considerably. If the animal is terrestrial, it may obtain water before ecdysis, since water may not be available in sufficient quantity at the time when the shell is shed.

The purple land crab, *Gecarcinus lateralis*, slowly takes up water, apparently through its gills, for several weeks before ecdysis and retains this water within its hemolymph. During the final hours before ecdysis, the old exoskeleton detaches completely and water may enter rapidly. This may become quite evident when the pericardial sacs, which are diverticula

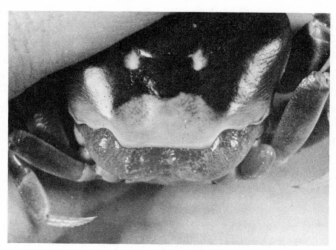

Fig. 100. Swollen pericardial sacs of purple land crab, shown just prior to ecdysis. Regenerating limb bud of fifth thoracic leg visible at right, lower rear of crab.

of the pericardium, swell, sometimes enormously. The pericardial sacs seem to be acting as hydrostatic equalizers. By their swelling, they provide more room within the hemocoel and thereby keep the hydrostatic pressure of the hemolymph more or less constant even as water enters.

For proecdysial uptake of water, the purple land crab is able to use dew and small puddles of rain water, or even water that lies between grains of thoroughly dampened beach sand or sandy soil. The crab may drink the water in rain puddles. More often, it presses the lower part of its body against wet soil or against drops of dew or rain water. With hair-like setae on its posterior ventral surface and with the wrinkled covering of its pericardial sacs, the crab draws the water up to its gills by capillary action, much as coffee is drawn upward through a lump of sugar when a corner of the lump is held in the coffee. The upward forces of capillarity may be supplemented by those of suction, created by the beating of the gill bailers within the crab's tightly sealed branchial chambers. According to American biologist Dr. Thomas G. Wolcott, the semiterrestrial ghost crab, *Ocypode quadrata*, by beating its gill bailers, can produce within sealed branchial chambers negative pressures of up to 76 mm below ambient pressure. These negative pressures can lift water from beach sand containing as little as three to five percent of water by weight.

During ecdysis in the purple land crab, first the gills, next the abdomen, soon the legs, and finally the claws are withdrawn from the old shell. Then the crab rids itself of the old chitinous lining of the foregut. If missing limbs have been regenerated, they are removed from small soft chitinous cases in which, twice folded, they have been formed. As hemolymph enters, the new limbs are stretched to full size, although they will not be as long as the other nonregenerated limbs for several more molts.

When finally out of its old shell, a land crab—or, for that matter, any shrimp, lobster, or crab—is soft, watery, and quite helpless, virtually unable to move, a prime target for predators. But in most cases the animal before ecdysis has taken refuge in a burrow, a rocky crevice, or some other

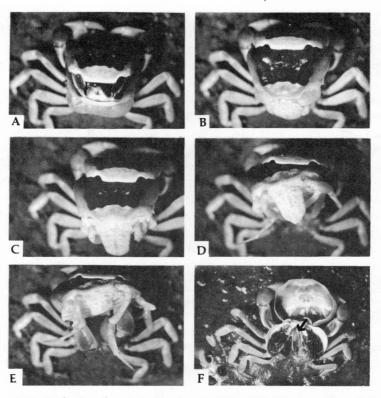

Fig. 101. Ecdysis of purple land crab (A)–(E). (F), interior of cast shell, showing lining of foregut (arrow) that was shed. (Photographs by author; reproduced through courtesy of the American Museum of Natural History)

hiding place. The animal remains secluded for several days following ecdysis, neither eating food nor moving about until its new shell is fairly hard. Then it starts emerging regularly from its retreat, seeking food and gradually resuming its normal pattern of activity.

An astounding aspect of molting is the apparent "rejuvenation" that seems to accompany it. Although before ecdysis the shell of a decapod crustacean may be battered, worn, cracked, faded, and otherwise in poor condition and the animal may lack vitality, after ecdysis the new shell is handsome, with bright pigments and healed wounds. And once the animal has hardened and begins to feed, it becomes very active, often

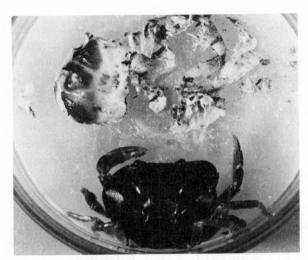

Fig. 102. Soft-shelled purple land crab a day or so after ecdysis, with remains of cast shell, which crab has been eating. Calcium salts in cast shell are used by crab to harden new shell.

much more so than prior to ecdysis. If molting were to continue indefinitely, a decapod crustacean would indeed seem to have discovered the fabled Fountain of Youth. But even a shrimp, lobster or crab must grow old. Some species of crabs undergo a so-called terminal ecdysis, which is followed by a frequently long period of time during which no molting occurs and which ends finally in the animal's death.

Whenever a shrimp, lobster or crab enters proecdysis, the animal undergoes marked changes in its physiology, and subsequent events occur, so to speak, automatically. The animal completes its preparations for ecdysis, sheds its old shell, enlarges and hardens its new shell, and then may enter anecdysis, the often prolonged nonmolting period. This terminates only when the animal again undergoes the physiological changes that accompany entrance into proecdysis. We may ask, what controls these physiological changes in the animal? What switches the animal from a condition characterized by no molting to one associated with molting?

The switch takes place when certain cells of the crustacean central nervous system, known because of their position and function as neurosecretory cells, withhold from the circulation a secretory product, a hormone that inhibits molting. There results within the hemolymph an increase in the concentration of a second hormone, the arthropod molting hormone known as crustecdysone. In crustaceans, this hormone—or, more likely, a precursor or inactive form —appears to come from the Y organs, which are paired glands in the anterior part of the thorax that must be present if a decapod crustacean is to molt. When Y organs are removed during anecdysis, the animal is blocked in early proecdysis. In ways that are not yet entirely clear, molting hormone causes in epithelial cells and in certain other cells some changes that constitute the first step in proecdysis. Presumably these changes induce other changes, and so the entire sequence of events known as molting is set in motion.

But what causes molt-inhibiting hormone to be withheld from the circulation? This depends partly upon the species of crustacean and its life style. In a freshwater crayfish, *Orconectes virilis*, for example, exposure to darkness may prevent the animal from molting, while exposure to light induces molting. From this, one might conclude that molt-inhibiting hormone is withheld from the circulation during exposure of the freshwater crayfish to light. In the purple land crab, environmental factors that favor survival at ecdysis, when the crab is soft and vulnerable, appear to cause the withholding of molt-inhibiting hormone. These factors are darkness, solitude, warmth, and moisture, the same factors that prevail at the bottom of a crab's burrow. Should environmental factors become unfavorable—should there be light, other crabs nearby, cold or heat, dryness—then molt-inhibiting hormone is released into the circulation and proecdysis is delayed for the period of time that the environment remains unfavorable.

The semi-terrestrial Indo-West Pacific crab *Ocypode macrocera* prepares for ecdysis when exposed to darkness, but also when exposed to light, provided the crab is on a light background, Dr. K. Ranga Rao has reported. This crab is light in color and by day and night runs freely on the beaches. Here its light coloration causes it to blend smoothly with the sandy background, thus protecting the crab from attack by predators. Should the crab wander by day from the beaches into the nearby woodlands and grasslands, its body would be silhouetted sharply against the dark soil, even though the crab is able to darken its body somewhat by dispersing the dark pigment within its chromatophores. A crab's ability to evade attack is slight when the animal is about to shed its shell, and on dark soil death would almost certainly follow. For this Indo-West Pacific crab to survive, therefore, molt-inhibiting hormone must be released into the circulation when the crab is exposed to light and on a dark background but is not needed when the crab is exposed to light and on a light background.

Through work by many scientists on a variety of vertebrates and invertebrates, there is now available an explanation of how an unfavorable environment can cause molt-inhibiting hormone to be released into the circulation. Apparently, information regarding environmental conditions is picked up by an animal's sensory receptors, converted into nerve impulses, channeled into the central nervous system, and transmitted to neurosecretory cells. The nerve impulses seem able to travel along the neurosecretory cell fibers to their endings, which are enlarged and filled with neurosecretory material that had been synthesized within the cell bodies and had migrated to the endings. When the nerve impulses reach the swollen endings, they cause material stored there to be released into the circulation. Included in this material is molt-inhibiting hormone.

Each stalked eye of a decapod crustacean contains some of the neurosecretory cells that synthesize and release molt-inhibiting hormone. In crabs, freshwater crayfishes, and at least some shrimps, but not necessarily in true lobsters or spiny lobsters, the surgical removal of both eyestalks, by removing a major source of molt-inhibiting hormone, generally induces molting. Lobsters may fail to molt on destalking, presumably because sufficient amounts of molt-inhibiting hormone may continue to be released from neurosecretory cells within the brain.

Even crabs, freshwater crayfishes, and shrimps may, under certain conditions, fail to molt following removal of eyestalks. The eyestalks of females, for example, contain, in addition to molt-inhibiting hormone, a hormone that inhibits the de-positing of yolk within eggs in the ovaries. When eyestalks are removed, yolk begins to be deposited, the ovaries en-large, and soon ripe eggs leave the ovaries, travel down the oviducts, and emerge, fertilized or unfertilized, from the genital openings.

Either molting or ovarian enlargement—but not both simultaneously—may follow destalking of a female crab, freshwater crayfish, or shrimp. What determines which takes place? Several scientists, including French biologist Dr. Noëlle Demeusy and Belgian biologist Dr. Adrien Bauchau, have concluded that the physiological process currently dom-inant in the animal is the one that occurs following destalking. As an example, in juvenile green crabs, *Carcinus maenas*, molting always follows eyestalk removal; but in mature adult green crabs, either male or female, development of the repro-ductive glands, or gonads, follows destalking. In other words, in a young, growing green crab, the dominant process is growth, while in a mature adult, it is reproduction.

Experiments by American biologist Dr. Mary Weitzman have shown that dominance of a physiological process may shift with the seasons. In the land crab *Gecarcinus lateralis*, a mature female destalked in spring undergoes ovarian de-velopment, but when destalked in autumn, the crab prepares for molting. Dr. T. Cheung, working at the University of Miami, found that the stone crab, *Menippe mercenaria*, also may respond to destalking at one time of year by molting, at another time by spawning.

Work of two other scientists, Dr. Rita Gomez Adiyodi of India and Dr. T. Ōtsu of Japan, suggested that in the brain and thoracic ganglionic mass of crabs there is a hormone that stimulates ovarian development even when the eyestalks are present. And this effect is enhanced if the Y organs, which appear to secrete molting hormone, or a precursor, are re-moved, Dr. Demeusy has shown. Thus, it would seem that ovarian development takes place when in the circulation there is less ovary-inhibiting hormone from the eyestalks and more ovary-stimulating hormone from the brain and thoracic ganglionic mass, while at the same time there is more molt-inhibiting hormone from the eyestalks and less molting hormone, or a precursor, from the Y organs.

As stated earlier, one process occurring during proecdysis is the regeneration of limbs that a crustacean may have lost since the previous ecdysis. In nature, such limb regeneration

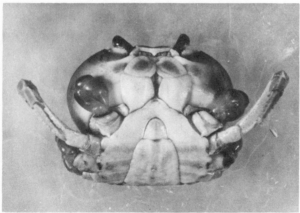

Fig. 103. Purple land crab one day before ecdysis, showing regenerating limb buds of claws and third through fifth thoracic legs. Top, dorsal view; bottom, ventral view. Loss of five or more limbs induces molting in this crab. (Photographs by author; top reproduced through courtesy of Zoological Institute, Lund, Sweden)

generally involves only one or two limbs, which, usually
through injury, have been reflexly detached at a plane near
the base of the limb known as the autotomy plane. In an intact
limb, a double membrane exists at this plane. Through the
double membrane pass a nerve and blood vessels that supply
nutrients to the limb. At autotomy, one portion of the double
membrane falls off with the limb, the other portion remains
with the stump, effectively sealing off the stump and virtually

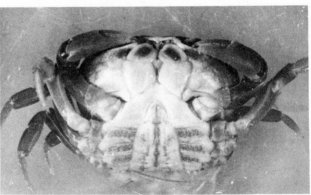

Fig. 104. Same crab as in preceding figure, but now soft, one day
after ecdysis. Top, dorsal view; bottom, ventral view. On left side,
third thoracic leg missing, probably injured and self-amputated
(autotomized) during ecdysis. Regenerated limbs smaller than
those nonregenerated.

eliminating bleeding. On the autotomy plane a new limb is regenerated, first as a tiny nipple-like structure known as a papilla, and subsequently, during the following proecdysis, as a twice-folded limb within a protective sac of soft chitin. The folded limb is extended to its full length during ecdysis. The new limb is smaller and often lighter in color than other, nonregenerated limbs, but otherwise it looks about the same.

In recent years, it has become clear, mainly through the work of Dr. Dorothy M. Skinner and her associates at the Oak Ridge National Laboratory, that, despite adverse environmental conditions, the loss of many limbs can induce a crab to molt. Such drastic loss of limbs can sometimes induce a female crab to molt when, if intact, she would be undergoing ovarian enlargement preparatory to spawning. The number of limbs that must be lost before molting is induced varies with the species of shrimp, lobster or crab. In the purple land crab, for instance, an animal can lose up to four limbs without effect, but loss of five or more limbs induces molting. Yet if a shrimp, lobster or crab is young and thus is already molting frequently, loss of many limbs may not further accelerate molting.

All of this experimental work has indicated how complex and delicate is the control of molting and gonadal development in decapod crustaceans and how important in exercising this control are neurosecretory components of the central nervous system, as well as various factors of the external and internal environment.

Eight

Sexual Differentiation

In vertebrates as in other animals, certain genes, which occur in chromosomes of cells, are responsible for the genetic sex of an animal. The genes determine whether the rudiment of sex organ, or gonad, is to develop into a functional testis or a functional ovary. The developing gonad of the vertebrate in turn secretes sex hormones that induce the development of secondary sex characters, thus making a genetic male appear masculine and a genetic female feminine. Nevertheless, rudiments of secondary sex characters typical of the opposite sex persist. Given an appropriate hormonal condition, the opposing secondary sex characters can supersede those of the genetic sex.

Sometimes, because of an abnormally developed or diseased gonad, an animal attains the so-called anhormonal (*an: without*) condition, in which the normal sex hormones are lacking. When in such a condition, the animal assumes the secondary sex characters of the anhormonal type. In birds, where the anhormonal type is predominantly masculine, female sex hormones secreted by a normally functioning ovary prevent genetically female birds from differentiating anhormonally as males. In mammals like ourselves, where the anhormonal type is predominantly feminine, male sex hormones from a normally functioning testis prevent genetic males from differentiating anhormonally as females. Yet occasionally the testes of a genetic male human being do not differentiate normally or, if they do so, they subsequently degenerate. As a result, the person becomes deficient in male sex hormones and gradually assumes anhormonal feminine secondary sex characters. A clinician may be well acquainted with examples of this type of human sex reversal.

In crustaceans the hormone that induces the development of male sex characters is secreted not by the testes, as in vertebrates, but by a pair of glands that are completely independent of the testes and are known as androgenic glands (literally "the glands that make the male").

In 1954 a French scientist, Dr. Hélène Charniaux-Cotton, brought to an end many decades of speculation regarding sexual differentiation in Crustacea by publishing the results of studies on an amphipod crustacean, the beach flea *Orchestia gammarella*. In her studies she discovered a pair of organs between muscles leading to the basal joint of the seventh thoracic legs. These organs she named the androgenic glands, because of the role that she found them to have in the determination of sex.

Dr. Charniaux-Cotton discovered that if she implanted an androgenic gland into either an immature female or a maturing female beach flea, the animal during subsequent molts gradually acquired male secondary sex characters. On its second thoracic legs a massive claw (used to hold the female during mating) developed, and the last (seventh) pair of thoracic legs became broader. Furthermore, within two or three months the ovaries of this animal had transformed into testes. When an androgenic gland was implanted into a mature female during her breeding period, deposition of yolk within her eggs did not take place.

Continuing her experiments, Dr. Charniaux-Cotton discovered that a female beach flea that had been masculinized by receiving an implant of androgenic gland proceeded to mate with a normal female that was about to lay eggs. Yet the

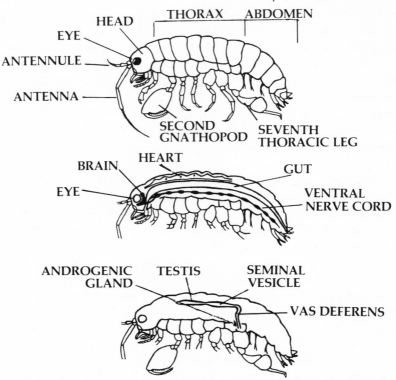

Fig. 105. Beach flea, *Orchestia gammarella,* **amphipod crustacean used extensively in research on sexual differentiation. Shown here, mature male, with enlarged second gnathopod and seventh thoracic leg. Top, external, lateral view; middle, internal, showing organs of digestive, circulatory, and nervous systems; bottom, internal, showing organs of male reproductive system.**

masculinized female could not fertilize the eggs, for in her sperm ducts, which had formed in response to androgenic hormone from the implant, the internal passageway, or lumen, through which the sperms normally would pass to the exterior, was not continuous with the gonad. Nevertheless, the sperms of the masculinized female were viable and, if removed from the gonads and placed on freshly laid eggs, were able to fertilize them. Subsequently, Dr. T. Ginsburger-Vogel, a co-worker of Dr. Charniaux-Cotton, obtained complete masculinization of female beach fleas by grafting androgenic glands into very young females, shortly after they had hatched. These masculinized females were able to fertilize eggs in the manner of normal males.

When Dr. Charniaux-Cotton implanted an ovary into a male beach flea that retained its androgenic glands, the implanted ovary rapidly transformed into a testis. In a male from which both androgenic glands had been removed, the male secondary sex characters began to regress and sperms no longer formed within the testes.

Some time before puberty, female crustaceans attain *permanent* sex characters. In *Orchestia gammarella* and other amphipods, these consist of brood plates, or oostegites (ō-ŎS-stĕ-gītes), which are flat projections of the second, third, and fourth thoracic limbs. Taken together, the oostegites comprise a brood pouch for the incubation of eggs.

At ecdysis just prior to egg-laying, female crustaceans acquire *temporary* sex characters, which, in the case of amphipods, are long hairs on the oostegites. In many, but not all, crustaceans, newly laid eggs become cemented to these hairs, which consequently are often called ovigerous, meaning "egg-bearing."

Dr. Charniaux-Cotton found that when ovaries were removed from a breeding female beach flea, the long ovigerous hairs were replaced by short juvenile hairs at the following ecdysis. Later, when the female from which ovaries had been removed received an implant of ovary, ovigerous hairs once more appeared on the oostegites—and at the same time yolk was deposited in eggs of the implanted ovary. When a male beach flea from which androgenic glands had been removed received an implant of ovary, yolk formed in the implanted ovary, and the male acquired oostegites and ovigerous hairs.

Unlike the testis, Dr. Charniaux-Cotton concluded, the crustacean ovary is an endocrine gland; that is, a gland that secretes hormones into the blood stream. In a young female beach flea, a hormone secreted by the ovary induces the formation of permanent sex characters, the oostegites. In older, mature females, what is presumably a second hormone from the ovary causes ovigerous hairs to appear at a time when yolk is being deposited in the ovary, in other words, just before egg-laying.

Thus, the differentiation of sex in the amphipod crustacean *Orchestia gammarella* requires, in the female, probably two hormones from the ovaries and, in the male, one hormone from the androgenic glands. To the present time, androgenic glands have been found in almost all groups of higher crustaceans, including isopods, amphipods, euphausiids, stomatopods, mysids, and decapods. It is likely, therefore, that in all of these groups the same kinds of hormones control sexual differentiation.

The androgenic glands appear to be controlled by neurohormones from the central nervous system. In several investigations, removal of eyestalks from the green crab, *Carcinus maenas*, and the freshwater crab *Potamon dehaani* hastened sexual development in immature males and caused exaggerated growth of the androgenic glands in both immature and mature males. Reimplantation of ganglia from the eyestalks into the freshwater crab prevented hypertrophy of the androgenic glands.

The anhormonal type in crustaceans is female. If androgenic glands are present and functional, there are two distinct effects: (1) the testes develop and (2) the male secondary sex characters develop. However, if the androgenic glands are not present or if they are present but not functioning properly, the ovary develops anhormonally and secretes hormones that lead to the development of female secondary sex characters.

Thus, the development of the testes requires that there be present in the circulation a particular hormone, the androgenic hormone. However, the development of the ovaries requires that androgenic hormone be absent from the circulation. If this requirement is met, the ovaries undergo autodifferentiation; that is, they develop spontaneously, without the intervention of a hormone. This was shown experimentally by Dr. Charniaux-Cotton. When she removed androgenic glands from certain male beach fleas, the testes transformed completely into ovaries.

Primordial female germ cells (oogonia) or the eggs (oocytes) to which they give rise may occasionally be found in testes of male crustaceans that retain their androgenic glands. In some way, perhaps through a decrease in the concentration of androgenic hormone, certain cells can escape the influence of the hormone and develop as female cells spontaneously, by autodifferentiation.

The possibility that oogonia and oocytes may develop within testes, despite the presence of androgenic hormone within the circulation, leads to a further possibility, that in nature there may exist intersexes among Crustacea. And indeed, intersexuality commonly occurs within the group. Some water fleas, or *Daphnia* (Branchiopoda), certain sow bugs (Isopoda), and some beach fleas like *Orchestia*, as well as other gammarid amphipods (those of the suborder Gammaridea) regularly exhibit intersexuality, wherein both male and female characters are present in the same animal. For instance, an individual may have in a single gonad both ovarian and testicular tissue. Or, of the two gonads, one may be an ovary and the other a testis. Or, as in the case of some gammarid amphipods, there may be both testes and a brood pouch.

Dr. Charniaux-Cotton has said she believes it likely that many, if not all, such cases of intersexuality may eventually be explained in terms of inadequately or abnormally functioning androgenic glands.

In subclass Cirripedia which includes the barnacles, there is a parasite known as *Sacculina*. After passing through an oceanic, free-swimming cypris larval stage, *Sacculina* discards most of its tissues and injects the remainder into a crab, where they attach to the ventral side of the intestine and send root-like branches throughout the host's body. From the root-like processes there appears a knob-like mass, which includes the rudiments of the reproductive organs and a ganglion. As it enlarges, the knob presses against the internal ventral surface of the host crab's abdomen and prevents the crab, at the ensuing molt, from forming any integument at this point. After the crab has shed its shell, the *Sacculina* can be seen projecting freely from the underside of the crab's abdomen. The projecting mass becomes a brood chamber for new *Sacculina*, with eggs developing into nauplii and hatching as free-swimming cypris larvae.

Male crabs that are infected by *Sacculina* become feminized. The abdomen gets broader and its margin more hairy. The first and second pairs of pleopods, which in the normal male crab are highly modified as copulatory organs, may approach the female type or degenerate altogether. Three posterior pairs of pleopods, which normally are lacking in the male, may develop.

Such changes in sacculinized male crabs were first described in 1886 by a French scientist, Dr. A. Giard. To sacculinization he applied the term parasitic castration, apparently because he believed that the primary action of the parasite was to destroy the gonads of the host. Subsequent workers found that gonads of parasitized crabs, although sometimes atrophied, remain functional. Occasionally testes of a parasitized crab contain oocytes or transform into ovaries.

Androgenic glands of a sacculinized male crab are not degenerate. The best explanation of sacculinization is that *Sacculina* removes androgenic hormone from the blood of the host and causes the concentration of the hormone in the host to fall. Because it requires a high level of this hormone for male secondary sex characters to be maintained, a gradual loss of male secondary sex characters, an appearance of oocytes in the testes, and a gradual assumption of female secondary sex characters result.

Sacculinization represents a type of sex reversal induced by parasitic infection. But there are numerous examples of spontaneous sex reversals in crustaceans, particularly in certain shrimps and prawns.

Among the more advanced shrimps, the Caridea, there are two superfamilies, the Alpheoidea and the Pandaloidea, in which many species exhibit sex reversal. Indeed, with one or two exceptions, all known species of pandalid shrimps spend the first portion of adult life as a male and the latter portion as a female.

For instance, according to Dr. T. H. Butler of the Fisheries Research Board of Canada, individuals of the arctic, deep-sea pink shrimp, *Pandalus borealis*, living in the Strait of Georgia along the east coast of Vancouver Island, British Columbia, hatch their eggs late in March and early in April, and the first two larval stages can be collected in plankton hauls in the inlet at that time. As the tiny young shrimps pass through four more larval stages, they migrate into shallower waters and there transform into juvenile shrimps. Within the next year, although a very small percentage may mature as true (genetic) females, most of the shrimps mature as males. At about two years of age the males enter a transitional stage, and by the time they are two and one-half years old, they are in their female phase. By the beginning of the winter they are carrying spawn, which hatch the following spring. The females may then live for another year before dying.

Thus, the pink shrimp may undergo a leisurely sex reversal, living for a time in a transitional phase after ceasing to be a functional male and before becoming a spawning female. During the transitional phase, secondary sex characters are intermediate between those of the male and female phases. Eight other species of pandalid shrimps occurring in the Strait of Georgia also go through a somewhat prolonged transitional phase during sex reversal.

Yet it is possible for a shrimp or prawn to undergo very rapid sex reversal. The English biologist Dr. David B. Carlisle reported that under certain favorable environmental conditions the shrimp *Lysmata seticaudata* was seen to copulate as a male, then 90 minutes later undergo ecdysis at which it lost the sex characters of the male and gained those of the female, and within an hour copulate again as a female.

Such change of sex part-way through adult life represents a type of hermaphroditism that is known as protandric. Simultaneous hermaphroditism, in which an individual can function both as a male and as a female simultaneously, is not known to occur among decapod crustaceans. Nor is alternating hermaphroditism, wherein an individual alternates from one sex to the other.

For the reversal of sex and the development of ovaries in a protandric hermaphrodite to occur, there must be a decrease in concentration of androgenic hormone within the circulation. According to Dr. Charniaux-Cotton, this decrease results from a degeneration of the androgenic glands, which takes place during the latter part of the male phase. Should such degeneration not take place, there would be no reversal of sex. Experiments on the shrimp *Pandalus platyceros* by American biologist Dr. Daniel L. Hoffman suggest that the degeneration of androgenic glands may be controlled by a hormone from the eyestalks. If eyestalks are removed from this shrimp in autumn, when reversal of sex normally takes place, the androgenic glands do not degenerate, the ovaries do not transform into testes, and there is no reversal of sex. A French biologist, Dr. A. Touir, found the same to be true in the shrimp *Lysmata seticaudata*.

Nine

Sea Farming

The term sea farming has been defined as the raising of marine organisms either from eggs or from more advanced stages of development to mature adults and then the harvesting of the adults for market. But, in the words of Dr. C. P. Idyll, author of the book *The Sea Against Hunger*, true sea farming, like farming on the land, "involves full control over the animal from birth to slaughter."

Nursery grounds for hosts of marine forms are the swampy areas on the edge of the sea known as marshlands, which are often mistakenly believed to be waste lands. These areas have tremendous potential in sea farming. But the artificial cultivation of living organisms in such areas can conceivably alter the environment almost as much as can pollution, siltation, interference with freshwater runoff, or direct exploitation. Thus, care must be taken that aquacultural techniques applied to these natural areas contribute to the efficient use of the areas and do not reduce their value for other purposes.

Among the most productive of marshlands are the mangrove swamps of Asia. According to Dr. Idyll, hundreds of pounds of protein food per acre annually come from fish farms in mangrove swamps of the Philippines, the Celebes, Java, Sumatra, and Taiwan, and along the Malabar Coast of India. He has compared the 300 to 1500 pounds of protein food per acre taken annually from fish ponds in some of these regions with the maximum of 800 pounds of beef per year that can be obtained with the most modern scientific techniques and best fertilizers from the best land available.

Some fish farms of the Orient are not, however, sea farms as defined by Dr. Idyll, for they omit the first step, the production of young from captured brood stock. Instead, the young are caught in the ocean or are trapped when washed into fish ponds by an in-flowing tide. In such ponds, so-called fish farming consists of the feeding and protection of young until they are large enough for market.

Aquatic farming is not a new development; it has been carried on for many years. Some freshwater fish, such as catfish, trout, and carp, are farmed; and for over a century marine fishes such as the cod, salmon and shad have been raised in fish hatcheries in the United States and abroad. Nonetheless, early marine fish hatcheries did not carry on true sea farming. Their purpose was to build up stocks of commercially desirable species by raising the young in a food-rich and predator-free environment and then turning the young out into the ocean when they reach a certain stage of development. Yet, except for two kinds of salmon, the coho and chinook, examination of catches failed to indicate an increase in their size after young fish were freed into the sea. Consequently, enthusiasm for fish hatcheries generally declined. A recent revival of interest, marked by plans to raise fish from hatching to marketable size, centers on a form of true sea farming and may meet with greater success than heretofore.

Sea farming of oysters has been carried on for a very long time. The ancient Romans cultivated oysters, and the Japanese have done this for centuries. Although crude by the standards of agriculture, oyster culture has advanced rapidly. Farming of oysters already accounts for most of the oyster harvest in many countries. Oyster farmers on the north shore of Long Island, New York, for instance, have maintained permanent brood stocks, which they have selected for higher rates of growth and better quality. They have induced oysters to spawn by manipulating temperature and then have maintained the young at optimal temperatures and in controlled culture media. After several weeks, during which the young would grow rapidly, the oyster farmers would transfer them, now known as spat, from culture rooms to the waters of Long Island Sound. Here precautions were taken to protect the spat from silting and predators. Such methods, developed for oyster farming, can be applied to the farming of scallops and clams.

A project designed to speed the harvesting of oysters, scallops and clams was undertaken at St. Croix in the Virgin Islands. Directed by Dr. Oswald Roels of the City University of New York, and the Lamont-Doherty Geological Observatory, Palisades, New York, and supported by a grant from the National Oceanic and Atmospheric Administration, the project involved pumping water from offshore at a depth of about 2900 feet into concrete ponds on the shore. Water brought into the ponds by this artificial form of upwelling, like that brought to the surface during natural upwelling, was cold, and rich in nutrients. Cultured diatoms (*Cyclotella nana, Chaetoceros simplex* and *Bellarochia)* in the ponds proliferated rapidly. They provided an abundant food supply for seed oysters, scallops and clams placed in the ponds after being shipped to St. Croix from laboratories on the mainland. The oysters, scallops and clams grew at unprecedented rates. Since costs are high, the potential of artificial upwelling in aquaculture may depend upon the applicability of this technique to other fields, such as the generation of electrical power, the production of fresh water, or air conditioning.

Among decapod crustaceans, shrimps seem generally best suited to sea farming, for they can grow rapidly while foraging and preying on various plants and animals near the bottom of the food pyramid, and they bring a good price in the markets. Commercial shrimp farms are currently operating in a number of tropical countries, and production is rising annually.

In very recent years, new techniques for inducing shrimps

to spawn in captivity have been developed and put into commercial use. Some shrimp farmers, nonetheless, use an older method in which gravid, or pregnant, female shrimps are purchased from trawl fishermen and allowed to spawn in large indoor tanks of running sea water. Eggs and larvae, which are free-swimming, are nurtured in the same tanks. Five to six weeks after spawning, the young shrimps, now living on the bottom, are moved to outdoor rearing ponds. Here, with supplemental feeding, they are fattened for market. Development from egg to marketable adult requires about seven months for shrimps spawned in spring and about 10 months for those spawned in autumn.

Farming of shrimps by this method was developed by a Japanese scientist, Dr. Motosaku Fujinaga, who subsequently became president of a company featuring the farming of the large kuruma shrimps (kuruma-ebi). These shrimps, *Penaeus japonicus,* are used in the preparation of the traditional Japanese dish tempura and must be kept alive until just before they are cooked. This places very fresh kuruma shrimps in great demand.

The Fujinaga method of culturing shrimps, mostly of the genera *Penaeus* and *Metapenaeus,* has been exported to other countries, such as Okinawa, Korea, Taiwan, France, Latin America and the United States. In recent modifications of the method that have greatly reduced costs, culture of larvae and postlarvae to advanced juveniles takes place entirely in outdoor concrete tanks.

Less sophisticated methods for raising shrimps are used in the Philippines and Taiwan, where juveniles of the large, highly valued species *Penaeus monodon,* known as the jumbo tiger shrimp, or "sugpo," are collected when they appear in inshore shallows. The young shrimps are taken to culture ponds and thence to rearing ponds, where they feed on algae, copepods and hosts of other forms and are protected from predators until large enough for market. This method has also been used commercially in Japan and experimentally in the United States. Capturing juvenile shrimps, however, is an uncertain procedure, often resulting in too small a catch for a fisherman's needs.

Within the United States, the raising of shrimps has been attempted commercially in only a very few places. Still, extensive research on shrimp culture is in progress. A dozen or so academic and governmental institutions have been involved in such research, and more than half a dozen industrial concerns have either worked actively on it or have provided financial support.

At the United States National Marine Fisheries Service Biological Laboratory in Galveston, Texas, some 50,000 penaeid shrimps can be hatched in a single culture tank and raised to postlarvae in about two weeks. The postlarvae are hardy and can be shipped by air in plastic bags, which contain sea water and supplemental oxygen and are packed in styrofoam boxes. When shrimp culture in the United States becomes commercially profitable, a few hatcheries may be able to supply many shrimp farms with postlarvae, to be raised to maturity. Or the postlarvae can be used to stock ponds, embayments, and inlets, where trawl fishermen make most of their catch.

In addition to penaeid shrimps, the freshwater shrimp *Macrobrachium* has also been an object of aquacultural research. Fisheries scientists in Malaysia have developed means of rearing, under controlled conditions, the long-legged and long-clawed giant freshwater prawn *Macrobrachium rosenbergii*. Although this prawn remains the favored species, rearing techniques have been adapted to other species of *Macrobrachium* at various laboratories around the world. In the Western Hemisphere, *M. carcinus* on the East Coast of the Americas and *M. americanum* on the West Coast have been studied for possible aquaculture at laboratories on the Gulf Coast of the United States and in Central and South America. Macrobrachium shrimps, though not yet abundant in the markets, are highly prized as specialty foods; when available, they compete favorably with penaeids.

According to Dr. Harold H. Webber, President of the Groton (Massachusetts) BioIndustries Development Company, although the culture of penaeids requires saline water and that of macrobrachium shrimps fresh water, the same hatchery (e.g., one situated on an estuary) may be used for both kinds of shrimps, since macrobrachium shrimps also need brackish water for egg and larval stages. Salinity of the water at the hatchery must be controlled. At such a hatchery, the temperature should remain between 75 and 90°F, for penaeids do not grow well at temperatures above and below those levels.

Geographically, Dr. Webber has said, these requirements tend to restrict shrimp farms to areas between the Tropic of Cancer and the Tropic of Capricorn. The Gulf of Mexico has large populations of shrimps, but the Gulf Coast of the United States is rather cold in winter for shore-based shrimp farming. After doing research on shrimp culture in the United States, some American firms have established their production facilities in other countries, where warm temperatures all year round and a more favorable economic climate make commercial production viable and more profitable.

Attempts to farm spiny lobsters and American lobsters, *Homarus americanus*, from eggs to adults of marketable size have not been commercially successful. Culturists of spiny lobsters, in particular, encounter great technical problems, mainly because these lobsters have a prolonged oceanic larval life of up to six or seven months. A possible approach is to capture young spiny lobsters after their planktonic larval life is complete and protect and feed them in enclosures until they are ready for market. In southern Japan attempts to do this have been successful; in the United States, the costs of such an operation may be excessive.

During the first half of the twentieth century, when fish hatcheries in the United States were experiencing their heyday, millions of American lobsters were raised to the fourth or fifth stage in lobster hatcheries of New England and New York and then released into the ocean. The object, of course, was to increase stocks of adult American lobsters. Unfortunately, so little benefit appeared to result from this attempt to "farm" lobsters that by the 1950s most lobster hatcheries were closed.

At the Massachusetts State Lobster Hatchery and Research Station, Oak Bluffs, Martha's Vineyard, however, culturing of American lobsters continued. Young lobsters were maintained for several weeks until they reached the fourth (post-larval) stage and then were released into coastal waters. Yet, even with constant care, survival to the fourth stage averaged less than 25% over a period of 11 years. Maximum survival in a single year was about 42%.

Attempts have been made at the Massachusetts State Lobster Hatchery and Research Station to speed the growth of young lobsters. By keeping the temperature of the water in the rearing tanks at 70°F all year round, scientists at the facility were able to bring the young to one-pound size in less than three years, approximately one-half the time that it takes to do this in nature. The scientists were interested in selective breeding for rapid rates of growth and for other desirable characteristics, including possibly two large crusher claws instead of one large crusher and a smaller cutter, or ripper, claw. The scientists also hoped to increase the number of red lobsters by crossing normal greens with reds. Obtaining more red lobsters and releasing them into coastal waters would facilitate population studies, for red lobsters are easier to track than are normal-colored ones.

A valuable by-product of research designed to make possible lobster farming is an increased understanding of the biology and behavior of American lobsters. This knowledge feeds back into the potential commercial operation, making profitable lobster farming all the more likely. According to John T. Hughes, Director of the Massachusetts State Lobster Hatchery and Research Station, there is a good chance that lobster farming will become a reality in the years ahead.

Farming of the American lobster's freshwater relative, the crayfish, has already become a reality. In Louisiana, "crawfishing" has long been an important industry that is centered about the red, or swamp, crayfish, *Procambarus clarkii,* and the white, or river, crayfish, *P. acutus acutus.* Within the past quarter century, these crayfishes have increasingly been farmed with simple techniques in wooded ponds, in open ponds often constructed solely to farm crayfish, and in rice fields, where crops of crayfish and rice are rotated. Management consists primarily of alternately flooding and draining the ponds and harvesting the crops. In

Louisiana alone, millions of pounds of crayfishes are harvested each year. Recently, fisheries scientists at Louisiana State University and elsewhere have found that these crayfishes can be farmed in brackish marshes, which in South Louisiana are extensive, provided the salinity of the water does not exceed eight to 10 parts per thousand.

Like the farming of American lobsters, the farming of commercially important crabs (e.g., blue crab, stone crab, Dungeness crab) on a commercial scale is not yet feasible. Nor is it likely to be for some time, due to the high costs of feeding and handling the animals during their development. Experimentally, however, progress has been made by scientists in several laboratories, notably by Drs. John H. Costlow and Cazlyn G. Bookhout of the Duke University Marine Laboratory, Beaufort, North Carolina. With the use of specially designed rearing facilities these scientists and their co-workers have been able to raise crabs of numerous species from egg to early adult stages.

On the whole, the outlook for mariculture of decapod crustaceans is not bleak. There is a real possibility that other economically important forms, in addition to shrimps, may eventually be farmed commercially. Through the use of molting hormones, it may be possible to speed the growth rate of shrimps, lobsters and crabs, particularly of the latter two, which require longer than do shrimps to pass through their developmental stages.

Arthropod molting hormones, known as ecdysones, are available. One form, α-ecdysone, was isolated in 1954 from silkworm pupae by West German scientists A. Butenandt and Peter Karlson and was synthesized in 1966 by American scientist J. B. Siddall and his co-workers at Syntex Research, Palo Alto, California. In 1966 Australian scientists F. Hampshire and D. H. S. Horn isolated another form, β-ecdysone (= crustecdysone, ecdysterone, 20-hydroxy-ecdysone) from spiny lobsters. Dr. Horn and other scientists have also obtained crustecdysone from plants and insects. It is now possible for research workers to purchase crystallized crustecdysone commercially under the name ecdysterone.

Crustecdysone is known to induce molting in some species

of insects and crustaceans. But this hormone does not induce molting in all of the species of crustaceans on which it has been tried. Scientists will have to experiment with various concentrations of the hormone and on different species of crustaceans before these advances in basic research can be applied to farming of shrimps, lobsters and crabs.

Should large-scale sea farming of shrimps, lobsters, and crabs become feasible, what benefits to society might result? According to Dr. Harold H. Webber, crustacean aquaculture will be characterized by high costs of production and high selling prices and thus will not be able to augment significantly the dietary protein needed by people in developing countries of Latin America, Africa, and Asia, where protein deficiency exists. Yet, because of favorably warm temperatures in waters of these areas, crustacean aquaculture may well be centered there. If so, it can contribute to the economic well-being of developing countries by providing employment and helping to generate a favorable balance of exchange.

In the highly developed industrial societies of Europe, North Africa, and Japan, on the other hand, according to Dr. Webber, a sizable market for shrimps, lobsters, and crabs already exists. Because of growing demand due to increasing affluence and improved standards of living, this market is unsatisfied and is limited by supply, with prices constantly on the rise. The supply of American lobsters available in the United States has declined, mainly because catches have remained virtually constant while imports have decreased. The supply of crabs also has decreased and that of spiny lobsters has changed relatively little. Furthermore, long-established trawling fisheries have almost completely exploited known natural populations of shrimps and prawns in oceans around the world. It is through crustacean aquaculture and sea farming that new populations of shrimps, lobsters and crabs must come, if significant increases in the supply of these seafoods are to develop.

Glossary

Anecdysis. A prolonged period without ecdysis; occurs in crustaceans that molt seasonally.

Anomuran crabs. Hermit crabs; coconut crab; also crablike members of the infraorder Anomura, in which the last (fifth) pair of thoracic legs is very small and often kept hidden within the branchial chambers.

Antennules. First pair of feelers, or antennae.

Anterior chamber. Second portion (after the esophagus) of the chitin-lined foregut; has often been called the cardiac stomach.

Aquaculture. Large-scale cultivation of aquatic organisms under controlled conditions, for use as food or for other economic purposes.

Brachyuran crabs (or true crabs). Members of the infraorder Brachyura, in which all five pairs of thoracic legs are well developed.

Branchial chambers. On either side of the cephalothorax of shrimps, lobsters, crabs and certain other crustaceans; formed from a deep lateral fold of the carapace; house the gills.

Carapace. The portion of the hard exoskeleton, or shell, that covers all or part of the body of many crustaceans; in shrimps, lobsters, and crabs, the carapace covers the head and thorax.

Caridean shrimps. Higher shrimps, members of the infraorder Caridea; include the majority of shrimps and prawns.

Cephalothorax. Region of the body in decapod crustaceans that is covered by the carapace, with the boundary between head and thorax indicated by the cervical groove. In lobsters the cephalothorax is called the "body;" in shrimps it is called the "head."

Chela. A claw, or the claw-like terminus of a cheliped.

Chelipeds. One or more pairs of thoracic legs of decapod crustaceans that terminate in a chela, or claw. Often the entire cheliped is referred to as a claw.

Chitin. A resistant complex chemical compound, the chief constituent of the exoskeleton, or shell, of crustaceans.

Class. A major subdivision of a phylum in the classification of animals, usually consisting of several subclasses and orders.

Crustaceans. Members of the superclass Crustacea of the phylum Arthropoda. Examples are shrimps, lobsters, crabs, amphipods, copepods, isopods, barnacles.

Decapod crustaceans. Members of the order Decapoda, class Malacostraca, superclass Crustacea, phylum Arthropoda; have five pairs of thoracic legs. Examples are shrimps, lobsters, crabs, hermit crabs.

Ecdysis. In crustaceans, the shedding of the old exoskeleton, or shell.

Family. A major subdivision of an order, suborder or infraorder in the classification of animals, usually consisting of several genera.

First-stage lobster. The first of three free-swimming larval stages of a true lobster; corresponds to the mysis stage of a penaeid shrimp.

Gastric mill. A food-grinding apparatus in the foregut of decapod crustaceans; in penaeid shrimps, consists of denticles (tooth-like processes) and lateral ridges in wall of anterior chamber; in crayfishes, lobsters and crabs, consists of one dorsal and two lateral teeth at junction of anterior and posterior chambers.

Genus (plural: *genera*). A main subdivision of a family or subfamily in the classification of animals, usually consisting of more than one species.

Gill bailer (or scaphognathite). Leaf-like flap in a channel at the anterior opening of each branchial chamber in decapod crustaceans; by its beating, drives water forward in the channel and out of the branchial chamber.

Glaucothoë. The postlarval stage of an anomuran crab.

Hard crab (or hard-shelled crab). Any crab that has not recently molted and is very hard, due to compounds of calcium and other minerals in the shell.

Hemocoel. The extensive spaces of an arthropod's body through which hemolymph circulates.

Hemolymph. The circulating and tissue-bathing fluid of arthropods; composed of cells and plasma; often loosely termed blood.

Infraorder. A major subdivision of a suborder in the classification of animals, usually consisting of several superfamilies and families.

Juvenile. A young shrimp, lobster or crab that has completed its larval and postlarval stages but is not yet a sexually mature adult.

Mandibles. Paired, jaw-like cutting, grinding or crushing appendages at the mouth of decapod and other crustaceans.

Mariculture. Large-scale cultivation of marine organisms under controlled conditions, for use as food or for other economic purposes; also utilization of the marine habitat for cultivation of aquatic organisms.

Megalopa (or megalops). The postlarval stage of a brachyuran crab.

Midgut glands (or hepatopancreas, or digestive glands). Paired glandular and storage organs of decapod crustaceans; digestive juices from midgut glands flow via tubules into the caudad (midgut) portion of the posterior chamber. Often called "fat," "liver," or tomally.

Molt. With reference to crustaceans, a clearly defined series of activities that precede, include and follow shedding of the shell (ecdysis). Among these activities are limb regeneration, decalcification of the old shell, laying down of the new shell, and hardening of the new shell.

Mucoid. Like mucus or related to mucus.

Mysis stage. The third in a series of free-swimming larval stages of penaeid shrimps.

Nauplius. The minute, egg-shaped or pear-shaped, earliest

larval form, into which many crustaceans, including penaeid shrimps, hatch from the egg. Higher shrimps, as well as lobsters and crabs, pass through the naupliar stage within the egg and hatch at a later stage of development.

Order. A major subdivision of a class in the classification of animals, usually consisting of several suborders and sometimes of infraorders.

Ossicles. Small, hard plates and projections of the internal gastric skeleton of decapod crustaceans; serve in part for attachment of muscles that move the foregut.

Ostia. Three pairs of small openings in the heart of decapod crustaceans; through the ostia, blood enters the heart from the pericardium.

Ovigerous. Bearing or carrying eggs.

Peeler crab. As used in the fishery for blue crabs, a crab about to shed its shell; recognizable by the red line along the edge of the swimming paddles (fifth pair of thoracic legs).

Penaeid shrimps. Primitive shrimps, members of the family Penaeidae; constitute the backbone of the shrimp fishing industry in North and South America and in many parts of Asia.

Pericardium (or pericardial sinus). The blood-filled area surrounding the heart of a decapod crustacean.

Petasma. Copulatory organ of male penaeid shrimps that is formed from the first abdominal appendages (pleopods); used by the male to thrust the spermatophore against the thelycum of the female.

Phyllosoma. The free-swimming larval stage of a spiny, or rock, lobster; corresponds in some ways to the protozoeal stage and in other ways to the mysis stage of a penaeid shrimp.

Phylum. A major primary subdivision of the animal kingdom, consisting of one or more related classes.

Plankton. Animal and plant life, largely microscopic, found floating and drifting in large numbers in the ocean and in bodies of fresh water.

Pleopods. Paired abdominal appendages used for swimming (hence often called swimmerets) by shrimps; used for attachment of eggs by female shrimps, lobsters and crabs.

Posterior chamber. Made up of the terminal portion of the

foregut and the anterior portion of the midgut; often called the pyloric stomach.

Proecdysis. The period of preparation for ecdysis.

Protozoea. The second in a series of free-swimming larval stages of a penaeid shrimp.

Puerulus. The postlarval stage of a spiny, or rock, lobster.

Seminal receptacle. In females of lobsters and some shrimps, an area on the ventral surface of the thorax that receives the spermatophore from the male during mating; in crabs, an enlarged portion of each oviduct that serves the same function.

Setae. Bristle-like, hair-like, or tooth-like processes on limbs and mouth parts of many crustaceans.

Soft crab (or soft-shelled crab). A crab that has just shed its shell and has not yet hardened its new shell; may be any species of crab, but the commercial soft-shelled crab of the Atlantic and Gulf coasts is the blue crab, *Callinectes sapidus.*

Spawning. Emission, or "laying," of eggs.

Species. A major subdivision of a genus or subgenus; regarded as the basic category of biological classification; composed of related individuals that resemble one another; are able (or potentially able) to breed among themselves, but are not able to breed with members of another species.

Spermatophore. A packet of sperms transferred by a male shrimp, lobster, or crab to the female during mating.

Spiny lobsters (or rock lobsters). Members of the family Palinuridae; lack large claws and have a flexible, leathery tail fan.

Sponge. The egg mass of a female crab, which she carries attached to long hairs on her pleopods.

Taxonomist. A scientist concerned with the classification of animals or plants.

Thelycum. On the ventral surface of the thorax of a female penaeid shrimp; receives the spermatophore from the male during mating.

Third maxillipeds. Outermost pair of mouth parts of a decapod crustacean; used, with the assistance of two other pairs of maxillipeds and two pairs of maxillae, for holding food until it can be pushed into the esophagus by the mandibles.

Thoracic legs. The five posterior pairs of thoracic appendages of a decapod crustacean; generally used for walking, although the first or second pair may be modified as large claws and the fifth pair as swimming paddles.

True lobsters. Members of the family Nephropidae; have large claws and a stiff tail fan.

Zoea. The free-swimming larval stage of anomuran or brachyuran (true) crabs; corresponds to the mysis stage of a penaeid shrimp..

Fishing Gear

Beam trawl. When small-sized, used to catch shrimps; consists of a net, the mouth of which is attached at its top to a heavy wooden beam; also a heavy ground rope that holds the lower edge of the mouth on the sea bottom; also two iron brackets (trawl heads) that support the beam and the sides of the net.

Crab dredge. As used in the winter fishery for hard female blue crabs, consists of a large iron frame to which a bag made of steel rings below and cotton twine above is attached, and a drag bar at the bottom of the frame, with long iron teeth that dig into the sea bottom and dislodge the crabs.

Crab pot. As used in the fishery for blue crabs, consists of a rigid metal frame about two feet square and covered with chicken wire; attracted by bait, crabs enter the lower section and swim upward, to be trapped in the upper section. Catches hard crabs, peeler females, and soft mating females carried by hard males. As used in the fishery for king crabs, consists of a frame about seven feet square made of welded steel bars and covered with stainless steel wire mesh or nylon netting. As used in the fishery for Dungeness crabs, is cylindrical, three to four feet in diameter, and made of iron or stainless steel rods covered with stainless steel wire mesh; usually has two funnel-shaped entrances.

Crab scrape. A large, toothless dredge used to catch soft blue crabs and peeler crabs; consists of a triangular metal frame to which a net is attached and which is pulled along the bottom by a small powerboat.

Crab trotline. As used in the fishery for blue crabs, consists of a long rope with a weight at each end and with bait tied at three- to five-foot intervals. As line is raised to surface, crabs clinging to bait are scooped into a net. Catches hard crabs, peeler females, and soft mating females carried by hard males.

Hoop net. As used in the South African fishery for rock lobster, consists of a deep conical bag supported on an iron ring and with several strings across the mouth to which bait is tied; used also to catch crabs in the Indo-West Pacific and lobsters and prawns in the British Isles.

Lobster pot. As used in the fishery for American lobster, consists of wooden laths nailed to a wooden frame shaped as a rectangle or half-cylinder; has two sections, the "kitchen," or "chamber," entered by the lobster seeking bait, and the "parlor," in which the lobster becomes entrapped. Modifications of this design also exist.

Otter trawl. As used in the fishery for shrimps and Norway lobsters, consists of a large, conical bag of netting or webbing, with a wide mouth and a cylindrical tail known as the cod-end; a line of corks, called the float line, that buoys up the top of the bag at its mouth; a leaded footrope or chain, called the lead line, that holds the lower part of the bag on the sea bottom; and heavy wooden boards, called otter boards, that are attached to the sides, or wings, of the bag. Float line and lead line are hung from the boards. As a motor trawler pulls the bag, the pressure of water against the otter boards keeps open the mouth of the bag and allows the catch to be scooped into the bag, where it collects in the cod-end.

Shedding float. As used in the fishery for blue crabs, is a large live car in which peeler crabs are kept until they have shed their shell; has a rough plank bottom and wooden laths on the sides, with a wooden flange along the outer sides that regulates the depth of submergence.

Select Reading List

Selection of titles to be included in a reading list is never an easy task. The 61 titles appearing here were culled from many hundreds in the scientific and popular literature. Recent major works, in some cases quite technical, were included if they provide ready information for the general reader and naturalist. In this category are checklists and keys, which frequently contain an extensive list of references. Titles of several bibliographies appear here as well.

Of outstanding value are the *FAO Species Identification Sheets for Fishery Purposes*, which, on a continuing basis, are being published by the Food and Agriculture Organization of the United Nations. Information sheets thus far published include those for crustaceans of the Mediterranean and Black Sea, the Western Central Atlantic, and the Eastern Central Atlantic; coverage around the world is planned. Distribution of the identification sheets is limited, but a reader can consult them in a specialized library, such as that of the American Museum of Natural History in New York City.

Of a general nature are Waldo Schmitt's *Crustaceans*, G. F. Warner's *The Biology of Crabs*, books by Cobb and by Cobb and Phillips (editors) on the biology of lobsters, and Iversen and Skinner's small volume on land hermit crabs "in nature and as pets." William Warner's Pulitzer prize-winning account of blue crabs and the watermen who fish for them in Chesapeake Bay is informative and a joy to read. For persons proficient in reading French, Bauchau's slim but informative and well-illustrated book, *La Vie des Crabes*, provides a summary of the European crab fauna and discusses the physiology of some forms.

For readers interested in fisheries, Doliber's *Lobstering Inshore and Offshore* provides extensive information. An entire issue of the *Marine Fisheries Review* (March–April, 1973) is concerned with shrimp fisheries, mainly in the Americas. Dumont and Sundstrom's account covers a wide variety of gear used in fisheries, including those for crustaceans.

Scientific and popular interest in culture of shrimps, lobsters, and crabs is running high. Fortunately, a recent book, highly recommended by a biologist familiar with the field, is available: Limburg's *Farming the Waters*.

For the benefit of readers whose interest in shrimps, lobsters and crabs is primarily gastronomic or that of a curious layman, included in this reading list are titles of popular articles and certain technical papers readily understood by an individual with little or no scientific background.

Some persons will undoubtedly wish to learn more about the intriguing animals discussed here. Delving into a few of the books and articles listed below or included within the bibliographies should provide an excellent start.

Adiyodi, K. G. and Rita G. Adiyodi. Endocrine Control of Reproduction in Decapod Crustacea. *Biological Reviews*, vol. 45, pp. 121–165, 1970.

Bauchau, A. G. *La Vie des Crabes*. 138 pp. Editions Paul Lechevalier, Paris. 1966.

Bliss, Dorothy E., Jacques van Montfrans, Margaret van Montfrans, and Jane R. Boyer. Behavior and Growth of the Land Crab *Gecarcinus lateralis* (Fréminville) in Southern Florida. *Bulletin of The American Museum of Natural History*, vol. 160, pp. 111–152, 1978.

Bliss, Dorothy E. From Sea to Tree: Saga of a Land Crab. *American Zoologist*, vol. 19, pp. 385–410, 1979.

Bright, Donald B. and Charles L. Hogue. A Synopsis of the Burrowing Land Crabs of the World and List of Their Arthropod Symbionts and Burrow Associates. *Los Angeles County Natural History Museum, Contributions in Science*, no. 220, 58 pp., 1972.

Chace, Fenner A., Jr. The Shrimps of the Smithsonian–Bredin Caribbean Expeditions with a Summary of the West Indian Shallow-Water Species (Crustacea: Decapoda: Natantia). *Smithsonian Contributions to Zoology,* no. 98, × + 179 pp., 1972.

Chace, Fenner A., Jr. and William H. Dumont. Spiny Lobsters—Identification, World Distribution and U.S. Trade. U.S. Fish and Wildlife Service, *Commercial Fisheries Review,* vol. 11, no. 5, pp. 1–12, 1949.

Chace, Fenner A., Jr. and Horton H. Hobbs, Jr. The Freshwater and Terrestrial Decapod Crustaceans of the West Indies with Special Reference to Dominica. *United States National Museum Bulletin,* no. 292, vi + 258 pp., 1969.

Charniaux-Cotton, Hélène. Androgenic Gland of Crustaceans. *General and Comparative Endocrinology,* Suppl. no. 1, pp. 241–247, 1962.

Cobb, J. Stanley. The American Lobster: The Biology of *Homarus americanus. University of Rhode Island, Zoology/ NOAA Sea Grant, Marine Technical Report,* no. 49, 32 pp., 1976.

Cobb, J. Stanley and Bruce F. Phillips (eds.) *The Biology and Management of Lobsters.* 2 vols. Academic Press, New York, 1980.

Crane, Jocelyn. *Fiddler Crabs of the World (Ocypodidae: Genus Uca).* 737 pp. Princeton University Press, Princeton, N.J., 1975.

Doliber, Earl L. *Lobstering Inshore and Offshore.* × + 108 pp. Association Press, New York, 1973.

Dumont, William H. and G. T. Sundstrom. Commercial Fishing Gear of the United States. U.S. Fish and Wildlife Service, Bureau of Commercial Fisheries, *Fish and Wildlife Circular,* no. 109, iv + 61 pp., 1961.

Felder, Darryl L. An Annotated Key to Crabs and Lobsters (Decapoda, Reptantia) from Coastal Waters of the Northwestern Gulf of Mexico. *Center for Wetland Resources, Louisiana State University, Baton Rouge, Publication* no. LSU-SG-73-02, viii + 103 pp., 1973.

Firger, Barbara A. A Lingering Look at Lobster Culture. *Oceans,* vol. 7, pp. 26–31, 1974.

Fischer, W. (ed.) *FAO Species Identification Sheets for Fishery*

Purposes. Food and Agriculture Organization of the United Nations, Rome. Mediterranean and Black Sea (Fishing Area 37), vol. 2, 1973 [includes crustaceans]. Western Central Atlantic (Fishing Area 31), vol. 6, 1978 [includes crustaceans].

Fischer, W., G. Bianchi, and W. B. Scott (eds.). *FAO Species Identification Sheets for Fishery Purposes*. Canada Funds-in-Trust, Department of Fisheries and Oceans, Canada, Ottawa, by arrangement with the Food and Agriculture Organization of the United Nations. Eastern Central Atlantic (Fishing Areas 34, 47 in part), vol. 6, 1981 [includes crustaceans].

Gladson, Jim. The Crayfish—Oregon's Freshwater Lobster. *Oregon Wildlife*, vol. 34, pp. 3–7, 1979.

Gray, George W., Jr., Robert S. Roys, Robert J. Simon, and Dexter F. Lall. Development of the King Crab Fishery off Kodiak Island. *Alaska Department of Fish and Game, Informational Leaflet*, no. 52, 16 pp., 1965.

Hardy, Sir Alister. *The Open Sea: Its Natural History*. One-volume edition. *Part I: The World of Plankton*. xvi + 335 pp. *Part II: Fish and Fisheries*. xiv + 322 pp. Houghton-Mifflin, Boston; The Riverside Press, Cambridge. 1965.

Hartnoll, R. G. Mating in the Brachyura. *Crustaceana*, vol. 16, pp. 161–181, 1969. (On mating in true crabs)

Hobbs, Horton H., Jr. A Checklist of the North and Middle American Crayfishes (Decapoda: Astacidae and Cambaridae). *Smithsonian Contributions to Zoology*, no. 166, iv + 161 pp., 1974.

Hobbs, Horton H., Jr. Synopsis of the Families and Genera of Crayfishes (Crustacea: Decapoda). *Smithsonian Contributions to Zoology*, no. 164, iv + 32 pp., 1974.

Holthuis, L. B. and H. Rosa, Jr. List of Species of Shrimps and Prawns of Economic Value. Food and Agriculture Organization of the United Nations, *FAO Fisheries Technical Paper*, no. 52, iii + 21 pp., 1965.

Idyll, Clarence P. *The Sea Against Hunger*. xiv + 221 pp. Thomas Y. Crowell, New York. 1970.

Idyll, Clarence P. The Crab That Shakes Hands. *National Geographic Magazine*, vol. 139, pp. 254–271, 1971. (About king crabs)

Ivanov, B. G. *A World Survey of the Shrimping Trade.* Moscow, 1964. Translated from the Russian by A. Mercado. Israel Program for Scientific Translations, Jerusalem. 1967. vi + 106 pp. Available from U.S. Department of Commerce, Clearinghouse for Federal Scientific and Technical Information, Springfield, Virginia 22151.

Iversen, E. S. The King-Sized Crab. *Sea Frontiers,* vol. 12, pp. 228–237, 1966.

Iversen, Edwin S. and Renate H. Skinner. *Land Hermit Crabs in Nature and as Pets.* 32 pp. Windward Publishing, Inc., Miami, 1977.

Kanciruk, Paul and Wiliam F. Herrnkind (eds.) An Indexed Bibliography of the Spiny Lobsters, Family Palinuridae. *State University of Florida, Sea Grant Program, Report* no. 8, 101 pp., 1976.

Lewis, R. D. A Bibliography of the Lobsters, Genus *Homarus.* *U.S. Fish and Wildlife Service, Special Scientific Report—Fisheries,* no. 591, 47 pp., 1970.

Limbaugh, Conrad, Harry Pederson, and Fenner A. Chace, Jr. Shrimps That Clean Fishes. *Bulletin of Marine Science of the Gulf and Caribbean,* vol. 11, pp. 237–257, 1961.

Limburg, Peter R. *Farming the Waters.* 223 pp. Beaufort Books, Inc., New York, Toronto. 1980.

Manning, Raymond B. and L. B. Holthuis. West African Brachyuran Crabs (Crustacea: Decapoda). *Smithsonian Contributions to Zoology,* no. 306, xii + 379 pp., 1981.

Marden, Luis. The American Lobster, Delectable Cannibal. *National Geographic Magazine,* vol. 143, pp. 462–487, 1973.

Pérez Farfante, Isabel. Western Atlantic Shrimps of the Genus *Penaeus.* U.S. Fish and Wildlife Service, Bureau of Commercial Fisheries, *Fishery Bulletin,* vol. 67, pp. 461–591, 1969.

Pickett, Joseph F., Sr., and Ronald J. Sloan. The Hidden World of the Crayfish. New York State Department of Environmental Conservation, *The Conservationist,* May–June, 1979, pp. 22–26.

Powell, Guy C. Growth of King Crabs in the Vicinity of Kodiak Island, Alaska. *Alaska Department of Fish and Game, Informational Leaflet,* no. 92, 106 pp., 1967.

Powell, Guy C. and Richard B. Nickerson. Reproduction of King Crabs, *Paralithodes camtschatica* (Tilesius). *Journal of the Fisheries Research Board of Canada,* vol. 22, pp. 101–111, 1965.

Powers, Lawrence W. A Catalogue and Bibliography to the Crabs (Brachyura) of the Gulf of Mexico. *Contributions in Marine Science,* vol. 20, Supplement, 190 pp., 1977.

Provenzano, Anthony J., Jr. The Shallow-Water Hermit Crabs of Florida. *Bulletin of Marine Science of the Gulf and Caribbean,* vol. 9, pp. 349–420, 1959.

Pyle, Robert and Eugene Cronin. The General Anatomy of the Blue Crab *Callinectes sapidus* Rathbun, *Maryland Department of Research and Education,* Publication no. 87, 40 pp., 1950.

Rees, George H. Edible Crabs of the United States. U.S. Fish and Wildlife Service, *Bureau of Commercial Fisheries, Fishery Leaflet,* no. 550, 18 pp., 1963.

Retamal, Marco A. Los Crustaceos Decapodos Chilenos de Importancia Economica. Universidad de Concepcion, Chile, Instituto de Biologia, *Gayana,* Zoologia, no. 39, 50 pp., 1977.

Roedel. Philip M. Shrimp '73—A Billion Dollar Business. *Marine Fisheries Review,* vol. 35, nos. 3–4, pp. 1, 2, 1973. (First of 21 articles on shrimps and shrimp fisheries, to which this entire issue is devoted.)

Schmitt, Waldo L. *Crustaceans.* 204 pp. University of Michigan Press, Ann Arbor. 1965.

Sims, Harold W., Jr. An Annotated Bibliography of the Spiny Lobsters—Families Palinuridae and Scyllaridae. *Florida Board of Conservation, Technical Series,* no. 48, v + 84 pp., 1966.

Stephens, William M. Land Crabs. *Sea Frontiers,* vol. 11, pp. 194–201, 1965.

Tagatz, Marlin E. and Ann Bowman Hall. Annotated Bibliography on the Fishing Industry and Biology of the Blue Crab, *Callinectes sapidus. National Marine Fisheries Service, Special Scientific Report—Fisheries Series* (NOAA Technical Report), no. 640, 94 pp., 1971.

Taissoun, Edgard. El Cangrejo de Tierra *Cardisoma guanhumi* (Latreille) en Venezuela: Distribución, Ecología, Biología y Evaluación Poblacional. *Boletin del Centro de Investigaciones*

Biológicas (Universidad del Zulia, Maracaibo, Venezuela), no. 10, 50 pp., 1974. (On great land crab)

Taissoun, Edgard. El Cangrejo de Tierra *Cardisoma guanhumi* (Latreille) en Venezuela: I. Metodos de Captura, Comercializacion e Industrializacion. II. Medidas y Recomendaciones para la Conservacion de la Especie. *Boletin del Centro de Investigaciones Biologicas,* no. 10, 36 pp., 1974.

Tinker, S. W. *Pacific Crustacea.* Charles E. Tuttle, Tokyo, 1965.

Van Engel, W. A. The Blue Crab and its Fishery in Chesapeake Bay. Part 1—Reproduction, Early Development, Growth, and Migration. U.S. Fish and Wildlife Service, *Commercial Fisheries Review,* vol. 20, pp. 6–17, 1958.

Warner, G. F. *The Biology of Crabs.* xvi + 202 pp. Van Nostrand Reinhold Company, New York; Paul Elek (Scientific Books) Ltd, London, 1977.

Warner, William W. *Beautiful Swimmers. Watermen, Crabs and the Chesapeake Bay.* xvi + 304 pp. Atlantic—Little, Brown, Boston, Toronto, 1976. Penguin Books, New York, 1977.

Williams, Austin B. Marine Decapod Crustaceans of the Carolinas. U.S. Fish and Wildlife Service, *Fishery Bulletin,* vol. 65, pp. 1–298, 1965.

Williams, Austin B. Marine Flora and Fauna of the Northeastern United States. Crustacea: Decapoda. *National Marine Fisheries Service, Circular* (NOAA Technical Report), no. 389, iv + 50 pp., 1974.

Wood, Carl E. Key to the Natantia (Crustacea, Decapoda) of the coastal waters on the Texas coast. *Contributions in Marine Science,* vol. 18, pp. 35–56, 1974. [Systematic key (with illustrations) for shallow water shrimp]

Young, Joseph H. Morphology of the White Shrimp *Penaeus setiferus* (Linnaeus 1758). U.S. Fish and Wildlife Service, *Fishery Bulletin,* vol. 59, no. 145, iv + 168 pp., 1959.

Zinn, Donald J. A Handbook for Beach Strollers. *University of Rhode Island, Marine Bulletin,* no. 12, iv + 116 pp., 1973. (Includes recipes)

Illustration Acknowledgments

Acknowledgment of Sources for Drawings

The author wishes to acknowledge with thanks the following sources, which provided significant artistic contributions to this book:

Fig. 1. *The Invertebrates*, by L. A. Borradaile, F. A. Potts, L. E. S. Eastham, and J. T. Saunders, 2nd. ed., 1955. Cambridge University Press, Cambridge, England.

Fig. 2. The biology of the shore crab, *Carcinus maenas* (L.), by J. H. Crothers, in *Field Studies*, vol. 2, no. 4, pp. 407–434, 1967. Publ. by Field Studies Council, Williton, Taunton, Somerset, England.

Fig. 3. Circulation and heart function, by Donald M. Maynard, in *The Physiology of Crustacea*, (Talbot H. Waterman, ed.), vol. 1, pp. 161–226, 1960. Academic Press, New York and London.

Fig. 5. *The Invertebrates: Function and Form*, by Irwin W. Sherman and Vilia G. Sherman, 1970. Macmillan Publishing Company, New York.

Fig. 8A. The shallow-water hermit crabs of Florida, by Anthony J. Provenzano, Jr., in *Bull. Marine Sci. of Gulf and Caribbean*, vol. 9, no. 4, pp. 349–420, 1959. (reprinted)

Fig. 8B. *Crustacea: Euphausiacea and Decapoda* (Fauna of the Clyde Sea Area), by J. A. Allen, 1967. Publ. by Scottish Marine Biological Association.

Fig. 10.   The benthic and pelagic habitats of the red crab, *Pleuroncodes planipes*, by Carl M. Boyd, in *Pacific Science*, vol. 21, no. 3, pp. 394–403, 1967.

Fig. 12.   *Crustaceans*, by Waldo L. Schmitt, 1965. University of Michigan Press, Ann Arbor.

Fig. 14.   (Same as for Fig. 10)

Fig. 20A. Lobsters: *Panulirus argus*, by Raymond B. Manning, in *FAO Species Identification Sheets for Fishery Purposes*, Western Central Atlantic (Fishing Area 31) (W. Fischer, ed.), vol. 6, 1978. Food and Agriculture Organization of the United Nations, Rome.

Fig. 24.   The northern shrimp fishery of Maine, by Leslie W. Scattergood, in U.S. Fish and Wildlife Service, *Commercial Fisheries Review*, vol. 14, no. 1, 16 pp., 1952.

Fig. 25.   Marine decapod crustaceans of the Carolinas, by Austin B. Williams, in U.S. Fish and Wildlife Service, *Fishery Bulletin*, vol. 65, no. 1, 298 pp., 1965.

Fig. 29.   Spiny lobsters—Identification, world distribution, and U.S. trade, by Fenner A. Chace, Jr. and William H. Dumont, in U.S. Fish and Wildlife Service, *Commercial Fisheries Review*, vol. 11, no. 5, 12 pp., 1949.

Fig. 31.   Slipper lobsters: *Scyllarides aequinoctialis*, by Raymond B. Manning, in *FAO Species Identification Sheets for Fishery Purposes*, Western Central Atlantic (Fishing Area 31) W. Fischer, ed., vol. 6, 1978. Food and Agriculture Organization of the U.N., Rome.

Fig. 33.   *The Life of Crustacea*, by W. T. Calman, 1911. Macmillan Publishing Company, New York.

Fig. 35.   Les crabes d'eau douce (Potamonidae), by Mary J. Rathbun, in *Nouvelles Archives du Museum d'Histoire Naturelle, Paris*, ser. 4, vol. 7, pp. 159–321, 1905.

Fig. 45.   A study of the reproductive organs of the common marine shrimp, *Penaeus setiferus* (Linnaeus), by Joseph E. King, in *Biological Bulletin*, vol. 94, pp. 244–262, 1948.

Fig. 60.   The general anatomy of the blue crab *Callinectes sapidus* Rathbun, by Robert W. Pyle and L. Eugene Cronin, in *Maryland* Board of Natural Resources,

Dept. of Research and Education, Publ. no. 87, 38 pp., 1950.

Fig. 65. (Same as for Fig. 60)

Fig. 66. Accouplement chez *Athanas nitescens* Leach, by H. and L. Nouvel, in *Bulletin de l'Institut Océanographique (Fondation Albert Ier, Prince de Monaco)*, no. 685, 1935.

Fig. 79. Aspects of social behavior in fiddler crabs, with special reference to *Uca maracoani* (Latreille), by Jocelyn Crane, in *Zoologica*, vol. 43, part 4, pp. 113–130, 1958. Publ. by New York Zoological Society. (reprinted)

Fig. 80. Biological and fishery research on Japanese king-crab *Paralithodes camtschatica* (Tilesius), by H. Marukawa, in *Journal of the Imperial Fisheries Experiment Station*, no. 4, 152 pp., 1933.

Fig. 81. *A Biology of Crustacea*, by James Green, 1961. Quadrangle Books, Chicago.

Fig. 82. The early life histories of some American Penaeidae, chiefly the commercial shrimp *Penaeus setiferus* (Linn.), by J. C. Pearson, in U.S. Dept, of Commerce, *Bureau of Fisheries, Bulletin*, vol. 49, no. 30, pp. 1–73, 1939.

Fig. 83. Natural history of the American lobster, by Francis H. Herrick, in U.S. *Bureau of Fisheries, Bulletin*, vol. 29, pp. 149–408, 1911. Also: The lobster fishery of the southern Gulf of St. Lawrence, by D. G. Wilder, in *Fisheries Research Board of Canada, General Series Circular* no. 24, 16 pp., 1954.

Fig. 84. The occurrence of phyllosomata off the Cape with particular reference to *Jasus lalandii*, by B. I. Lazarus, in *South Africa Division of Sea Fisheries, Investigational Report* no. 63, 38 pp., 1967.

Fig. 86. (Same as for Fig. 80)

Fig. 90. The biology and conservation of of the blue crab, by Curtis L. Newcombe, in *Virginia Fisheries Laboratory of the College of William and Mary and Commission of Fisheries, Educational Series* no. 3, 15 pp., 1943. Also: Experimental sponge-crab plantings and crab larvae distribution in the region of Crisfield, Md., by James G. Graham and G. Francis Beaven, in *Maryland Dept. of Research and Education, Publ.* no. 52, 18 pp., 1942. Also: The blue crab in Maryland estuarine waters, by D. G. Cargo, in *Maryland Tidewater News*, vol. 12, no. 2, Supplement no. 6, 1955. (Maryland Dept. of Research and Education)

Fig. 97. Environmental regulation of growth in the decapod crustacean *Gecarcinus lateralis*, by Dorothy E. Bliss and Jane Rouillion Boyer, in *General and Comparative Endocrinology*, vol. 4, no. 1, pp. 15–41, 1964. (© Academic Press, New York)

Fig. 105. Sex determination, by Hélène Charniaux-Cotton, in *The Physiology of Crustacea* (Talbot H. Waterman, ed.), vol. 1, pp. 411–447, 1960. Academic Press, New York and London.

Most of the artwork appearing here was prepared by Cady Goldfield. Others contributing drawings were Jane Boyer, Anthony J. Provenzano, Jr., Julie C. Emsley, and the author.

Acknowledgment of Sources of Photographs

The author wishes to acknowledge with thanks permission to use the following photographs in this book:

Fig. 6. Courtesy of Talbot H. Waterman and K. W. Horch, *Science*, vol. 154, pp. 467–475, 1966. Copyright 1966 by the American Association for the Advancement of Science.

Fig. 7. Courtesy of the American Museum of Natural History.

Fig. 9. Courtesy of Ralph Buchsbaum.

Fig. 13. Copyright by Douglas Faulkner.

Fig. 15. Courtesy of Wakefield/Pacific Pearl Seafoods.

Fig. 17. Copyright by Bob and Ira Spring.

Fig. 21. Courtesy of the Virginia Institute of Marine Science, Gloucester Point, Virginia.

Fig. 22. (Same as for Fig. 21)

Fig. 26.   Courtesy of the Maine Department of Marine Resources.

Fig. 27.   Courtesy of the National Geographic Society. © National Geographic Society.

Fig. 34.   Courtesy of *The Florida Naturalist*, October, 1976, vol. 49, no. 5. Florida Audubon Society, Maitland, Florida. © 1976. Photograph by author.

Fig. 36.   Courtesy of the *American Zoologist*. Photograph by author.

Fig. 37.   Courtesy of the American Museum of Natural History. Photograph by author.

Fig. 38.   Courtesy of Jacques van Montfrans.

Fig. 57.   Courtesy of the New England Aquarium. Photograph by Mary Price.

Fig. 67.   Courtesy of P. F. Berry, *Crustaceana*, vol. 17, pp. 223–224, 1969. © E. J. Brill, Leiden.

Fig. 68.   Courtesy of the National Geographic Society. Photograph by Robert Sisson. © National Geographic Society.

Figs. 69–71.   Courtesy of the Virginia Institute of Marine Science, Gloucester Point, Virginia.

Figs. 72, 73.   Courtesy of C. Dale Snow, *Journal of the Fisheries Research Board of Canada*, vol. 23, pp. 1319–1323, 1966.

Figs. 74, 75.   Courtesy of Eric Edwards and the Ministry of Agriculture, Fisheries, and Food, London.

Figs. 76, 77.   Courtesy of Eric Edwards and the Ministry of Agriculture, Fisheries, and Food, London; also *Crustaceana*, vol. 10, pp. 23–30, 1966. © E. J. Brill, Leiden.

Fig. 78.   Courtesy of the American Museum of Natural History. Photograph by Jacques van Montfrans.

Fig. 85.   Courtesy of the American Museum of Natural History.

Fig. 87.   Courtesy of the National Geographic Society. Photograph by Robert Sisson. © National Geographic Society.

Fig. 88.   Courtesy of the National Geographic Society.

Photograph by Robert Sisson. © National Geographic Society.

Fig. 89. Courtesy of the American Museum of Natural History.

Fig. 91. Courtesy of the Virginia Institute of Marine Science, Gloucester Point, Virginia.

Fig. 92. Courtesy of Jack Stark.

Fig. 93. Courtesy of the American Museum of Natural History. Photograph by Jacques van Montfrans.

Fig. 94. Courtesy of the American Museum of Natural History. Photograph by Jacques van Montfrans.

Fig. 95B. Courtesy of the *American Zoologist*. Photograph by author.

Fig. 96. Courtesy of the American Museum of Natural History. Photograph by Jacques van Montfrans.

Fig. 98. Courtesy of the Zoological Institute, Lund, Sweden. Photograph by author.

Fig. 99. Courtesy of the *American Zoologist*. Photograph by Frank White.

Fig. 101. Courtesy of the American Museum of Natural History. Photographs by author.

Fig. 103 (top). Courtesy of the Zoological Institute, Lund, Sweden. Photograph by author.

The following photographs were taken by the author (some have already been noted above but are included here as well). Figs. 34, 36, 37, 42, 46A, 46B, 48–56, 58, 59, 61, 62, 64, 95A, 95B, 98, 100–104.

Index of Animals
(Crustaceans)

Includes scientific name and author, popular or descriptive name, and group(s). Major groups of crustaceans appear in Figure 4.

Decapods comprise shrimps—mainly penaeid and caridean
lobsters—spiny and true (plus freshwater crayfishes)
crabs—anomuran and true (brachyuran)

General Index